Digitally prepared, printed and bound in Canada

Design by Amstier Graphics Inc.

National Library of Canada Cataloguing in Publication Data

James, Melbourne V.

Make the wind blow

ISBN 0-9694247-8-7

1. Wind tunnels. 2. Aerodynamics–Research–Canada.

I. Title

TL567.W5J34 2001 629.134'52 C2001-902939-X

The author and publisher have endeavoured as far as possible to ensure the accuracy of the content and information in this book. If there are any inaccuracies, please inform either of the above so that the correction may be made in any subsequent editions.

MAKE THE WIND BLOW

by

Melbourne V. James

HEIRLOOM PUBLISHING INC.

Table of Contents

Preface

Creating wind to test many of the things we take for granted is what this book is all about. We fly in planes, we drive automobiles or ride buses and trains, we cross bridges, we live or work in high-rise buildings, all without giving thought that there are people who test whether or not these vehicles or structures are safe. There are also those who create such chambers to aid wind engineers in making sure they are secure. This book is about both groups and, in particular, about Canada's leading role over the past 50 years in creating, worldwide, numerous aerodynamic, climatic, acoustic and other wind tunnels to test aeroplanes, vehicles, buildings, bridges – even the gas and oils we use in our automobiles – to enable us to enjoy greater mileage, safer road conditions, and quieter rides for driver and passenger comfort. It's been a fascinating experience to talk not only to the people involved in such work every day, but also to the founders of DSMA who made Canada a recognized world leader in wind tunnel design and construction, contributed to our use of nuclear power and played a key role in the development of the Canadarm that is, today, helping build a station in space.

As writer, I am indebted to some 50 people who allowed me to interview them, some of the retired members sharing fond memories of their days at DSMA ATCON and DSMA International. Others helped patiently guide me through the pitfalls of not being an engineer capable of fully understanding their profession. I can't include every name in this latter group, but I would be totally remiss if I did not mention Dr. Gary Elfstrom and Albert Bullock of Aiolos Engineering and the three members of the original company, Paul Dilworth, Lloyd Secord and George Meagher. I also wish to acknowledge the work of Peter Reitsma of Amstier Graphics, for the design and layout of the book, and Angela Dea Cappelli Clark, Helen de Verteuil and Charles J. Humber of Heirloom Publishing Inc. And no list would be complete without mentioning my wife Barbara, who edited the first draft and was a calming influence when I raged that the computer wasn't doing what I asked of it or I misplaced papers on my cluttered desk.

Mel James

1 Creating Wind for a Train

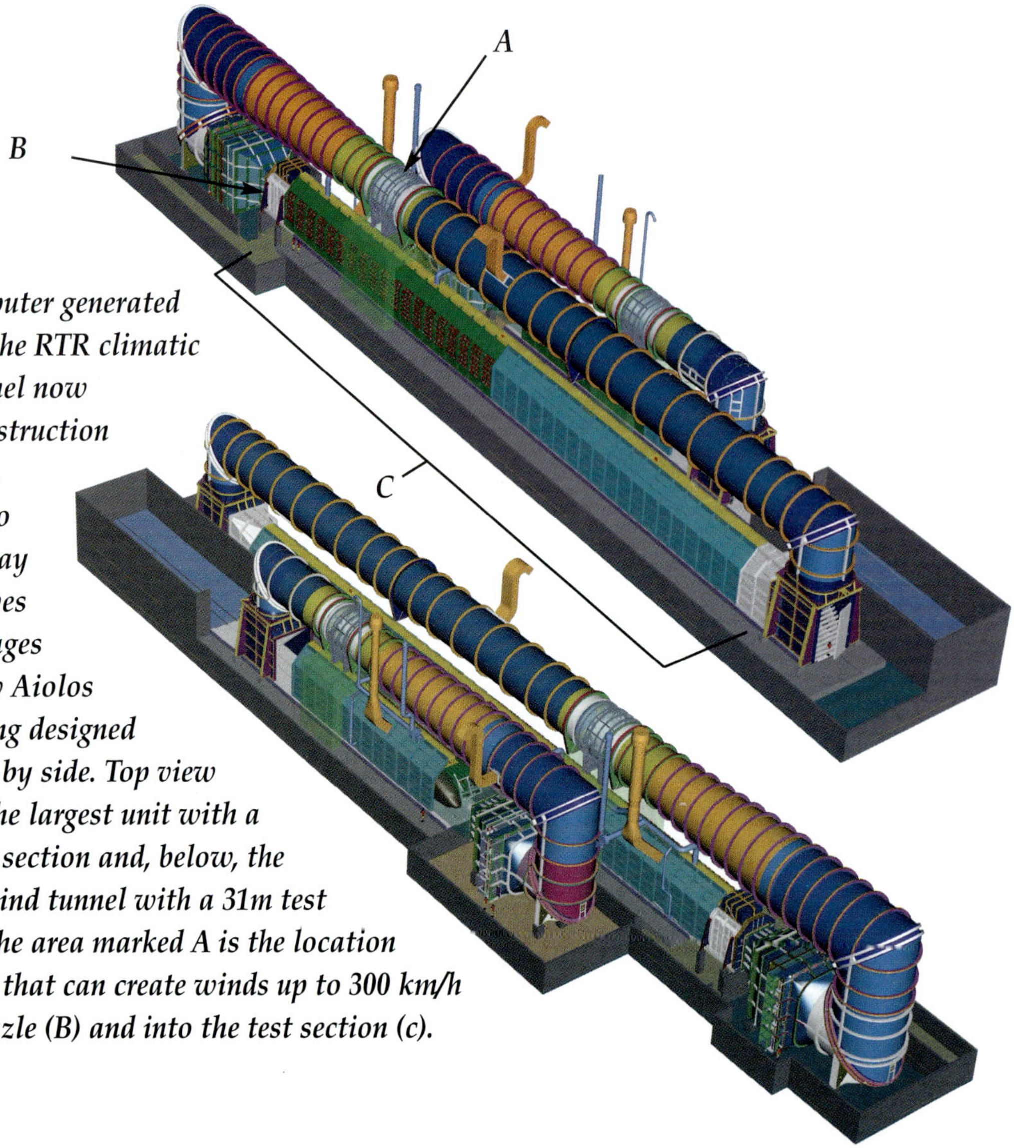

Two computer generated views of the RTR climatic wind tunnel now under construction in Vienna, Austria, to test railway locomotives and carriages show how Aiolos Engineering designed them side by side. Top view features the largest unit with a 100m test section and, below, the smaller wind tunnel with a 31m test section. The area marked A is the location of the fan that can create winds up to 300 km/h to the nozzle (B) and into the test section (c).

On July 11, 2000, Gerrit O. Zorgdrager, vice-president and general manager, and Adrian Hirsch, as manager of contracts for Aiolos Engineering Company, signed a contract with Austria's Federal Minister of Transport, Michael Schmid, to build what will become when its commissioning takes place late in 2002, the world's longest climatic wind tunnel. The signing ceremony was held at the Palais Schwarzenberg in Vienna with the city's Councillor of Finance, Dr. Brigitte Ederer, also present at the ceremony that launched what a press release described as, "a worldwide unique test plant."

It is unique for several reasons. It is one of the few climatic wind tunnels in the world "designed especially to test rail vehicles (locomotives, rail cars, subway cars, trams)," the release explains, "as well as road and special purpose vehicles." It also allows the simulation of extreme climatic conditions of between "minus 50 to plus 60 degrees Celsius, and provides rain, snow, and solar radiation with wind speeds up to 300 km/h."

To the staff of Aiolos Engineering Corporation who spent months designing specific plans to meet the requirements of the client (a consortium of European rail manufacturers and the Austrian government), it's the latest

The nozzle to regulate the air flow into the test section is a key unit in wind tunnel design and Aiolos Engineering created one of its largest to serve the RTR project. Its opening measures 4.6 X 5.3 metres.

proof that Canada is one of the world's front runners when it comes to designing and constructing wind tunnels.

It is a reputation Canada has enjoyed since a company known as Dilworth, Secord, Meagher and Associates (DSMA) worked closely with the Federal government's National Research Council to design a high speed wind tunnel at Ottawa's Uplands Airport in the late 1950s, a low speed one in the 1960s and since then has designed and/or built tunnels to test aircraft and automobiles in at least 17 countries around the world. This includes the USA, India, Italy, Holland, Germany, England, Sweden, Japan, Korea, Romania and Yugoslavia. At the present time, Aiolos is involved in the design and/or building of at least four other tunnel projects in the USA, Korea, and France.

Clients have included national governments, major auto makers, oil companies, and universities to teach aerodynamics and what is more commonly identified as "wind engineering." The University of Western Ontario which built a wind tunnel in the mid-1960s and collaborated with DSMA to build a second one in the 1980s, not only uses it to teach wind engineering, but also, under the direction of Professor Alan Davenpor,t carries out numerous tests on buildings and bridges around the world. Other universities have had tunnels built by DSMA and now Aiolos, the latest being at the Universiti Teknologi, Malaysia, fully operational and handed over in May, 2001.

RWDI, a private company in Guelph, Ontario, a city located roughly half way between the University of Western Ontario and Toronto where DSMA and Aiolos got their start, is also a major player in providing wind tunnel testing for a worldwide client list that has included the Academy of Firearms of the FBI in the United States, the Skarnsund bridge of Norway and the 101-storey Taipei Financial Center in Taiwan. RWDI projects have also included retractable roof studies of a number of sports complexes, including, in one case, an assessment of how the wind affects the trajectory of a baseball when hit.

Wind tunnels (or wind engineering) has become a major Canadian export with the contract in Austria being one of the largest climatic testing facilities in the transportation industry. Actually, Aiolos recommended that two wind tunnels be created within a single building, the complex to be roughly one and a half times the size of a Canadian football field. They did this based on the size of the site itself, which was somewhat limited, and the fact that it would actually cost less than the original specifications demanded. The revised amount is an estimated 59.6 million Euro dollars – 8.7 million of which will come to Aiolos as the design and project engineers. It was also realized that two tunnels would provide greater flexibility as it would be possible to carry out two different tests at the same time.

The two tunnels will replace a smaller 40-year old wind tunnel that has "reached the end of its technical period of use." The new facility will not only test external weather conditions, but also assess the interior of the train's heating, ventilation and air conditioning systems, a relatively new demand in the international certification process carried out in Europe where rail equipment is assessed on the basis of where it will be used.

Trains serving northern Europe, for instance, will undergo tests to assess extremely frigid conditions that can occur in Siberia, Finland and other rail lines operating above the 60th parallel. The climatic wind tunnel can be cooled to minus 50 degrees, and will test such things as the brakes and windshield wipers of locomotives and the freeze-up of doors and other components of cars and engines in extreme weather conditions. How vehicles will react to snow and icy conditions is assessed for equipment serving the northern European countries, while those serving the more southern climes will be tested and certified after they have undergone temperature tests above the 40 plus level Celsius, where a low temperature test might be assessed at the minus 10 degree level. Trains travelling between Switzerland and Italy, for example, can experience everything from icy conditions to extreme heat and can experience other unusual

A technician installs one of the solar lamps in the 100m test section of the RTR project. The bank of lamps will make it possible to simulate the intense heat of the sun on the passenger carriages. Another area of the test section will simulate ice and snow conditions with temperatures dropping as low as minus 50 degrees Celsius.

conditions while travelling through long tunnels. All the rail equipment provided by the Canadian company Bombardier operating in Europe (for the "Chunnel" between England and France, for example,) underwent a battery of tests reflecting tunnel conditions before being approved.

In Europe, therefore, tests of rail equipment must meet certain criteria for different regions of the continent and this can only be economically accomplished by wind tunnel testing – a development which is relatively new as a technology or science. Certainly sea captains and sail makers of old concerned themselves with wind conditions but didn't build wind tunnels. No one seems quite sure who created the first one but it is known that Gustave Eiffel tested his tower in a wind tunnel of his own design before it was erected on the Champ de Mars for the Universal Exposition of 1869. It is said, in fact, that some of the tests for Montreal's Olympic Stadium underwent tests in the tunnel built by Eiffel, and it is certainly known that the Wright Brothers tested models in a six foot long wind tunnel of their own design before they flew at Kitty Hawk, N.C. on December 17, 1903. In Canada, Wallace Rupert Turnbull of Rothsey, N.B., whose invention of the variable pitch propeller in 1927 revolutionized the aircraft industry at that time, is credited, at the turn of the century, with building the first wind tunnel out of an old packing case.

All, of course, were crude by today's standards, as was the first one built by the National Research Council (NRC) in an old factory on John Street in Ottawa in the 1930s. During the war two others were built and are still in use in the NRC complex on Montreal Road. In fact, an article in the October 2000 edition of *Canadian Cycle* magazine reports on tests carried out on a number of motorcycles in the 2m x 3m low speed wind tunnel in the basement of the building, with Kevin Cooper, NRC's Principal Research Officer, assessing the tests results of various models.

When the National Aeronautical Establishment of the NRC won government approval to build a high speed wind tunnel at the Ottawa Uplands airport in 1957, they sought outside help for the engineering and construction of the project. Dilworth, Secord and Associates, which later became known as Dilworth, Secord Meagher and Associates (DSMA), a consulting engineering firm founded in 1952 by Canada's pioneer jet engine developer Paul B. Dilworth, won the contract in 1957 and went on to become recognized as one of the leading wind tunnel designers in the world. Besides the high aerodynamic speed facilities, such as the one at Uplands, DSMA also developed wind tunnels for testing aircraft, automobiles, buildings and bridges under various climatic conditions. Others have been built to assess radiators and the reaction of gasoline and oil products under various weather and driving conditions. More recently, acoustic wind tunnels have been in demand in response to the growing demand for reduced noise levels in aeroplanes, automobiles and industrial plants around the world, and wind tunnels are the most economical means of determining what is required.

While some wind tunnels are described as "aerodynamic," others are referred to as "climatic" or "acoustic." All of them are designed to test the concerns of a particular manufacturer. Another type, described as "boundary layer" wind tunnels, are mostly used by architects, designers and structural engineers of buildings and bridges. Their interest includes determining the cladding, glass thickness and frame strength for windows, the effect of snow on roofs or the sway of a bridge in normal and storm conditions.

Some wind tunnels can produce winds up to 15 times the speed of sound, or Mach 15. Others meet the needs of their customers by providing tunnels that produce winds of 15 feet per second, or 60 km/h.

Most high speed aeronautical wind tunnels cannot economically be built or operated to accommodate actual (full scale) vehicles and so sub-scale models are used. A non-dimensional number, the Reynolds number, is used to characterize how faithful the flow around and, especially, close to, an aircraft surface is replicated using a scale model. Typically, the Reynolds number is too small in a wind tunnel test, but this shortfall can be compensated by increasing the air density through use of higher pressure or lower temperature. Examples of relatively high Reynolds number facilities include the Pilot European Transonic Wind Tunnel, which operates down to minus 150 degrees Celsius, and the NRC high speed wind tunnel at Uplands, which can create pressures up to 10 times normal atmosphere. Both of these high tech facilities were engineered by DSMA staff.

There are also water flume wind tunnels to assess the erosion of shore lines, study the requirements of off-shore oil rigs, the effect of fishing nets being lowered off the coast of Newfoundland, or to assess the effect of the sails used for various yacht races. Three such tunnels are at RWDI in Guelph, the University of Western Ontario, and Memorial University in St. John's, Newfoundland.

Because of the special requirements of each client, just about every wind tunnel built in the last 40 years is individually designed, and DSMA, and now Aiolos, have been involved in at least 40% of them worldwide. The trisonic or "blowdown" wind tunnel at Ottawa with a test section measuring 1.5m X 1.5m square can produce wind speeds of Mach 4.2 or 5100 km/h and has been used to test models of military and civilian aircraft for the past 40 years. Other blowdown tunnels designed or built by DSMA are located in India, Romania and Yugoslavia.

Low speed tunnels for aircraft and automobile testing started with the wind tunnel at Ottawa in the 1960s and, since then, design and/or building contracts have been completed for Chrysler in the USA, Ford in Germany, England and USA, General Motors in England and USA, Porsche and Mercedes Benz of Germany, Saab and Volvo of Sweden, Hyundai of Korea, Honda and Toyota of Japan and VAZ/Lada of Russia. Some are used for aerodynamic tests, and nearly all include a capability for air climatic testing. Wind tunnels have also been designed and built to test racing cars for Williams Grand Prix of England, the All American Racers of California and another in Italy to test Ferrari racers.

Some car manufacturers want facilities to test the actual vehicles while others develop accurately designed car models to test in smaller wind tunnels. A good example of a full scale unit is the NRC low speed wind tunnel in Ottawa which has a test section 9m x 9m. (The test section cross section size is the common denominator used to describe the size of a wind tunnel.) Ford-Werke AG in Germany had DSMA provide engineering services for wind tunnels to test automobiles built from 1/6 scale to full scale. One of the smallest wind tunnels is also at the NRC in Ottawa. Built in the 1940s to test components of various kinds, it is still in use. Its test section is a mere eight inches square.

Besides vehicle aerodynamics, wind tunnels are used in many areas such as providing different climatic conditions or acoustical needs. Boundary layer tunnels, for example, are usually built to a scale of one to 400 to measure wind velocities on tall buildings, models of which are built and set up in the test section. The CN tower in Toronto, for example, tested by the University of Western Ontario, was represented by a model measuring just over three feet tall. Tests on this model revealed that at 1136 feet, which is the highest level for sightseeing outdoors, wind velocities could be at least 30% greater than ground level. An assessment of numerous bridges at both the RWDI and the University of Western Ontario have also established that not only wind but rain and icy conditions on the cables of suspension bridges can have a major effect on the stability of the structure.

Wind tunnels have also been used to help scientifically determine failures in design of bridges and other structures, such as the collapse of the Tacoma Bridge in 1940 and the sinking of the oil rig *Ocean Ranger* off the shores of the Grand Banks of Newfoundland on February 15, 1982, with the loss of 84 lives. In the Royal Commission hearings that followed, the NRC was commissioned and gave evidence based on what they learned by submitting a scale model in its water flume tunnel to assess why the rig had capsized.

Wind tunnels, therefore, play a major role in determining safety factors, in reducing gas and oil consumption, in helping provide comfort inside aeroplanes, trains and automobiles and in numerous other ways. Canada has been a world leader in developing and testing the effect of such chambers for more than 40 years.

2 Dilworth, Secord, Meagher and Associates – The Early Years

Paul Dilworth, second from the left, shares a few memories with employees of Aiolos Engineering Company, a number of them former members of his own company, Dilworth, Secord, Meagher and Associates, before his retirement in 1978. This year 2000 photo, taken in the Aiolos board room, shows, left to right, Sergio Raimondo (back to camera), Paul Dilworth, Graham Thomas, Carl Balkwill, Tony Ferreira, Ervin Kass, Michael Higgins, Albert Bullock, Dan Simec, and Roy Fedoruk sitting at the table.

The founding of Dilworth, Secord, Meagher and Associates was entirely due to one of Canada's jet engine pioneers, Paul B. Dilworth. His interest and aptitude in engines and flying was evident from his early teens, first by taking apart the engine of the family motorboat to see how it worked and, while still in his teens, winning the Ellis Trophy for flying proficiency at the Toronto Flying Club in 1934 while attending the University of Toronto. Graduating in mechanical engineering in 1939, he joined the engine laboratory of the National Research Council and, in 1942, was selected as the junior member of a three-man team to go to England and report back on how Canada might best assist Great Britain's wartime jet engine development.

The team, headed by Charles A. Banks and including Ken Tupper, then head of the National Research Council's Hydraulics Laboratories, suggested that a cold weather station be established in Canada in order to verify performance and compressibility limitations of jet engines at low temperatures encountered in high altitude flight. Stevenson Field in Winnipeg was selected as the location for Ken and Paul to set up the Cold Weather Test Site (CWTS). On January 4, 1944, they carried out their first test – the first time the whine of a jet engine was heard in Canada.

This led to the establishment of Turbo Research Limited, a crown corporation, with Tupper named chief engineer to carry out research, design and development of gas turbine engines, while Paul remained in Winnipeg at CWTS until 1946. That year, the Canadian Government decided to transfer all work on gas turbine engines to private industry with A.V. Roe Canada Limited and Paul, who earlier in the year was transferred to Turbo Research, became chief engineer of A.V. Roe's gas turbine division.

Supported by a remarkable team of young engineers, the first Chinook axial flow engine with a 2600 lb thrust was running by March 1948. This led to a contract to develop a 6500 lb thrust engine for the *CF100* twin-engine jet fighter. Designated TR5, and later named Orenda, the engine later set a world speed record and, by 1957, nearly 4000 of them were manufactured not only for Canadian aircraft but also were sold to air forces in The Netherlands, Belgium, West Germany and South Africa.

Paul, however, decided to quit in 1952 as manager and chief engineer of what was simply known as the engine division of A.V. Roe, to form Paul Dilworth & Company. He wasn't sure what direction his company should take – buy a small manufacturing business in which new products could be developed, establish a straightforward consulting engineering practice, or make sophisticated products for small and medium sized companies which could not support an engineering department of their own.

His first step was to call on Walter Gordon of the Clarkson Gordon Chartered Accountant firm where his father had been a partner. "Gordon," recalls Paul, "admitted he could offer little advice relating to the business aspects of my concept, but made some helpful general suggestions on setting up the company." He also contacted Brigadier F.C. Wallace, a former boss at Turbo Research, who had become a senior officer of Duplate Canada, part of the Argus Corporation. Paul recalls he was "cautiously supportive of my plan," but warned that "while the need for the service was very real, most small and medium sized firms lacked the financial resources to pay for any major product development programs."

Despite the warning, Paul decided to launch the firm and rented space in a non-descript small manufacturing building at 1197 Lawrence Avenue West, in Toronto. While Paul decided he would not "lure any of my old colleagues to join me," he already had, by the time the office opened, two with him. Lloyd C. Secord, a gold medal winner in mechanical engineering from Queens in 1942, was one of the first hired at Turbo Research. Secord had already become chief designer at A.V. Roe, but left following a reorganization of the company in 1953, and called on Paul. "I was more than happy to have him join me and offered him a post as an associate," Paul recalls.

Another to share the office with Paul was George Kelk, in charge of instrument development at A.V. Roe, who also decided to leave A.V. Roe as a result of the reorganization. Paul remembers, "we decided on a loose association whereby we would share office space and George, whose technical genius was clearly evident, would serve as an associate consultant." Another ex-Roe employee, Dick Shepherd, a professor at Cornell University, "also agreed to serve as an associate consultant," wrote Paul in his brief notes about the founding of the company.

The first job came from DeHavilland Canada Limited when Phil Garratt, its managing director, encouraged the head of his engineering department to steer some design work to Paul's fledgling enterprise. It was to develop a castered tail wheel for DeHavilland's new single engine Otter aircraft and Joe Purvis, one of the eight engineering graduates recruited for Turbo Research in 1945, spent evenings and weekends helping develop the wheel that became "our first revenue producing contract."

Lloyd, meanwhile, had contacted the Continental Engine Company of Detroit. They offered him employment there but he suggested he should be a consultant instead and worked on a line of small gas turbine engines they were developing under a license agreement for Turbomecca of France. Paul recalls, "They were delighted to give him a retainer but Continental had some military contracts which made it necessary to secure mandatory security clearances through Washington."

At the time," Paul adds, "Senator Joseph McCarthy was raising a furor over communist infiltration so the application got pigeonholed for months until I happened to mention it to Clint Spurr, of Wheeling, West Virginia, a fellow cottager on Peninsula Lake in the Muskokas. He said he knew the secretary of defense and would have a word with him, and Lloyd soon became associated with the Continental company for more than a decade." Paul, given one job of stress analysis on a turbine blade by Continental, recalls ruefully, "I had to dust off years of neglect of my engineering education to do the job." He also recalls they were paid in U.S. currency, when the Canadian dollar was at a premium of five percent over the U.S. dollar. "What a difference that would have made today," he laughs.

In 1956, the company was incorporated as Dilworth Secord and Associates Ltd., and moved to a commercial rather than an industrial district of Toronto's west end – 4210 Dundas Street. Another "fallout" from A.V. Roe joined them: an Irishman named Matt Ryan "who resigned from A.V. Roe in a flare of Irish temper. He was a bright engineer and I was glad to take him on," Paul wrote in his autobiographical notes.

Besides Paul and Lloyd, a host of bright, creative engineers joined the firm over the years. Yet one of their first major jobs turned out to be a disaster. The Carrier Air Conditioning Corporation of the United States, in Syracuse, N.Y., gave them a job to design, develop and test a high powered fan for the air conditioning system for the new Douglas *DC6* aircraft. "We decided to design the fan on the 'Hero' principle using hollow fan blades through which hot air bled from the engine was inducted at the blade root and ejected through a trailing edge orifice at the blade tip," Paul wrote. "It was a neat concept which we saw as a great reputation builder for our fledgling firm. Instead it was an unmitigated disaster."

The fan would not come up to full speed because the design had failed to allow for the energy required to impart the spin entailed getting the inlet air up to the tip speed of the blades. Before a solution was found the unit suffered several blade failures due to a mathematical error that had been made in calculating the centrifugal loading of the blade root.

"These gross errors resulted in the termination of our contract with Carrier and they made no response to my letter of apology," Paul remembers. As a result, however, a rigorous practice of independent checking on both engineering calculations and final drawings was instituted, and the firm went on to build an enviable reputation on other Canadian and foreign contracts.

In 1954, Paul and Lloyd launched their first subsidiary, Champlain Power Products (CPP), because, as Paul Dilworth explains, "We didn't want the consulting firm caught up in the hardware business." Champlain Power Products was created when it became known that the *CF 105 Arrow* program would require a ground power unit (GPU) to supply both electric and compressed air power and supply of air conditioned air to operate over a temperature range from tropical to arctic conditions. It was an expertise Paul had as the former chief of the cold weather station CWTS at Stevenson Field in Winnipeg and as chief engineer of the gas turbine division at A.V. Roe (Avro) before leaving them to open his own company.

"We had anticipated that Avro might retain a consulting firm to design and engineer this GPU unit and that they would handle the task of getting the units built and ready to deliver to the RCAF," Paul wrote. "The RCAF, however, decided to handle the whole project themselves and called for 'turnkey' design, manufacture and supply at a fixed price.

We then set up CPP to bid on the turnkey job, with Brigadier Wallace prepared through one of his companies to provide the necessary financing."

Paul and Lloyd, whose expertise was well known at A.V. Roe, weren't known to officials of the RCAF and they awarded the contract to others. "As it turned out, however, the whole *Arrow* and *Iroquois* program was cancelled by decree of the Diefenbaker government," Paul's memorandum continues, "including the GPU and all other work relating to that program."

One of the next major challenges was solving a track ventilation problem at the CNR's Central Station in Montreal. The station was originally designed so that steam engines used for passenger trains would be replaced by electric locomotives about half a mile from the station, so the trains could be pulled into or out of the station located under a multi-storey building. The system worked fine until diesels replaced the steam locomotives. Realizing that diesels could bring the trains into the station and that the switching of engines was costly and time consuming, the question was raised why diesels shouldn't replace the electric locomotives. There was one major problem – smoke and exhaust fumes from the diesels would cause discomfort, not only for passengers leaving or entering the trains but also those working in the buildings built over the track area. One engineering firm's proposal to mount large fans to blow across the whole station track area was rejected as the wind thus created was too high for passenger comfort, particularly in the winter months. This prompted Dr. Omond Solandt, vice-president of Research for the CNR, to decide he wanted to get a separate opinion from some unconventional engineers he knew, so invited Paul and Lloyd for the task.

They accepted and handed over the task to Dr. Ian J. Billington, an M.A. Sc., graduate from the University of British Columbia (1952) and the first Ph.D graduate in aeronautical engineering hired by DSMA in 1957. He proposed a ceiling exhaust extraction system to remove the diesel exhaust that would capture most of the locomotive exhaust gas as it was discharged and transport it through the ducts to carry it well beyond the station and the surrounding buildings. "We were obliged to warn Dr. Solandt that there would still be some problems in the track area when a locomotive engineer occasionally yanked the throttle open and created a belch of excessive smoke," Paul recalls, "but Dr. Solandt said he felt they could live with that and we were given the green light to go ahead." One small section of the track area was done first to make sure the duct proposal would work. It did and all the remaining roof area was covered with the ducts and extended as other buildings were added over the track area, such as the building of the Bonaventure Hotel in 1966.

The entry into the design and building of wind tunnels that was to make DSMA one of the world's most recognized and respected wind tunnel firms in the world took place in 1956. That year the National Aeronautical Establishment (NAE) decided to build a trisonic blowdown wind tunnel at Uplands Airport, to be jointly operated by the National Research Council and the Defense Research Board. The decision to build it was partly prompted by the ongoing research and development of the Avro *Arrow* supersonic jet interceptor being designed by another former colleague of Paul's, Jim Floyd. Paul decided to make a pitch for the job.

It was a huge undertaking as the specifications called for a wind tunnel with a five by five foot test section (originally called the "5ft. x 5ft." but with the introduction of metric in the 1970s is now known as the "1.5m X 1.5m" test section), through which wind speeds from low subsonic levels through the "sonic barrier" up to supersonic speeds as high as Mach 4.2

could be obtained. To make this possible an industrial air compressor of 10,000hp would be required to draw the air from outside, and filter, dry compress and store it in three interconnected tanks, each 11.5 feet in diameter and 157 feet long. A control valve was designed to totally release the air from the tanks in times ranging from 6 to 20 seconds. The test section itself would have the capacity to operate with either solid or perforated walls, the latter when testing around the sonic (Mach 1) level. A model support system called a sting allowed pitch, yaw and roll orientations of the object being tested.

Paul, recognizing that he was unlikely to win such a major contract for a firm that consisted of five or six people at the time, teamed up with his ex-associate, Ken Tupper, who has been one of the three sent to England in 1942. Tupper had just left as dean of the University of Toronto's Faculty of Applied Science to become head of Ewbank and Partners Canada Ltd., a branch of a highly successful firm of civil engineering consultants in Great Britain.

With the Ewbank Dilworth Secord team in place, Paul approached the Defence Research Board but ran into trouble when he met with another colleague of NRC days, Dr. J.J. "Johnny" Green. He had become the Defence Research Board's senior research office and, as such, expressed the view that "no Canadian firm would be given the job since none had the necessary experience or expertise and Canada could not justify the expense of any Canadian firm having to learn it from scratch."

Paul countered that he would employ the services of a U.S. firm with the necessary specialist skills and that the potential for subsequent overseas customers for wind tunnels would more than justify the learning costs entailed. Green, however, was not convinced and the meeting ended in an impasse. Dilworth and Tupper, however, weren't about to let the matter drop. They contacted Dean MacKenzie, president of the National Research Council and the Honourable C. D. Howe, then Minister of Trade and Commerce. Perhaps even more important a contact was made when they talked to another old boss and Canada's most highly regarded pioneer of aeronautical engineering, J.B. Parkin, who was director of the NRC's Mechanical Engineering Division and directly involved in the building of the wind tunnel for the Defense Research Board.

"An ever ardent Canadian," Paul recalls, "Parkin, when informed of J.J. Green's statement, made an equally vehement policy statement to the effect that it should be assigned to a Canadian firm and had pounded the table to emphatically drive home his views."

To bolster their case, Ken visited Sverdrup and Purcell, a large engineering firm in St. Louis, Missouri, that had engineered the continuous flow 16-foot supersonic wind tunnel for the U.S. Airforce Arnold Engineering Development Center at Tullahoma, Tennessee. "Eventually the contract was awarded to our two firms, Dilworth Secord and Dilworth-Ewbank, the latter informally formed as a special arrangement for the project."

"We accepted the job under a cost–plus contract and justified the miserly profit fee of $10,000 on a multi-million dollar project on the grounds that the work would enable both our new firms to build staff and gain experience and reputation." This was made more amenable when the contract manager agreed to let Ken and Paul bill their personal time on the job as he wanted to ensure that the project got proper direction from the principals of the two firms. J.A. Parkin also helped by making it possible for them to hire Don Nasser from NRC who had done the preliminary deign work under Julius Lukasiewicz, the NRC's chief research engineer who had made the conceptual drawings. Nasser became project manager.

According to Uldis Kiesners, who joined the Dilworth-Ewbank team in August 1957, Paul Dilworth played a key role in attracting a number of other bright versatile staff to the project, many of them later becoming long-term career employees of the company Dilworth, Secord, Meagher and Associates and subsequently, DSMA International or DSMA ATCON. Among them were Tony Stonehill who worked on the aerodynamic design, Peter Tillson on controls, Alfred Bauer on the fixed diffuser and exhaust section. Uldis, a native of Latvia, worked on the design of the variable diffuser and transonic test section. Another brilliant member to join the team at that time was Ernst Eggmann, a native of Switzerland newly emigrated to Canada, who was hired even though he couldn't speak English. He could, however, communicate things through Dr. Jap Deleuw at the University of Toronto Institute of Aerospace Studies (UTIAS) who was working part time for Dilworth-Ewbank. In this way Ernie devised a superb design of the sting model support system, one of the key components influencing the overall performance of the wind tunnel, and over the next two decades, Ernie's designs of model support structures, fans, telescopes and other projects were considered "genius" by many staff members, most of them recalling a favourite anecdote about working with him.

With the cancellation of the *Arrow* in 1958, there was a concern that the high speed wind tunnel would suffer the same fate, and project activity was in fact curtailed while the government decided whether or not to complete the contract. Paul learned later that the project went ahead on the strength of an argument put up by another former NRC colleague, John Orr, who had become a senior official in the Department of Industry and argued that the experience would lead to overseas contracts. "The startling news of the Russian launch of *Sputnik* also awakened the unscientific world of the impact of new technology in the field of aerodynamics and this undoubtedly conditioned the thinking of bureaucrats and politicians," Lloyd adds.

Another NRC person to become deeply involved was Herb Kelland, a native of Chilliwack, B.C., and a 1941 graduate of the University of British Columbia who joined the NRC in 1943 and moved to its newly opened flight research centre in 1946. Ten years later, he acted as a liaison officer for the Defense Research Board that was providing the funding for the wind tunnel, a post that deeply involved him in the various contracts awarded. "There were a wide variety of bids from the contractors," he recalls. "A contractor who might bid the lowest for one component was often the highest for another aspect of the project and highest on another element was the lowest in something else." Coupled with that was the initial cost estimate, "a guesstimate really," in Herb's words, "that the tunnel could be built for $2 million when an even more realistic doubling of that figure was exceeded." The final expenditure for the building, wind tunnel structure and other tools required amounted to $9 million, "but if you were to build it today you'd have to increase that by at least a factor of eight," Herb declares.

Eventually seven consultants and 11 contractors were selected, and despite the close co-operation and expertise of the National Research Council, the project encountered a few major problems. When it was discovered that the contractor failed to install specified foundation reinforcement, the architect insisted the whole structure must be torn down. Another architect, however, found a way to install the necessary reinforcement without having to raze the structure. It was also thought they had another major problem when it was discovered that the complex bolt structure imbedded in the concrete base for the air compressor and motor was misplaced. Harry Ewbank, however, on a visit from England, dismissed the problem with a wave of the hand and the comment, "Oh … that happens all the time."

Despite these problems, and to reduce further cost overruns, the NRC decided, when the wind tunnel was completed in 1961, to commission it themselves and gave the job to a relatively newcomer to its ranks, Bill Rainbird. A native of New Zealand, Bill left a teaching and consulting career as an aerodynamicist in England to join the Defense Research Board in 1958, and was later assigned to the NRC to help oversee the construction of the tunnel with Herb Kelland. Bill was then called on to use his expertise to calibrate and commission the project.

The trisonic blowdown wind tunnel at Uplands Airport has been in operation ever since. Now 40 years later, companies such as Bombardier and other manufacturers of both civilian and military aircraft around the world mount their models on its various support systems to measure the reaction of them in wind velocities as low as 300 feet per second or up to Mach 4.2. It was the beginning of a long and close relationship between the NRC and Dilworth, Secord, which had become Dilworth, Secord, Meagher and Associates in 1958. That's when George Meagher joined the company as a senior principal partner and promoted the business so the company became recognized as one of the most admired and respected engineering consulting firms in the world.

Besides its expertise in designing and building wind tunnels, DSMA became involved in almost every major telescope project worldwide. It also teamed up with Atomic Energy of Canada Limited (AECL) and Ontario Hydro to play a leading role in nuclear energy projects in

Canada and became prime consultant for the world's largest cyclotron at the University of British Columbia. It played a major role in the testing and development of the Canadarm for the NASA space shuttle program and gained further recognition at home and abroad for hundreds of other jobs that many other engineering firms considered beyond their capability.

"It was our policy to take on virtually any task, and give our engineers the freedom to solve the problem using sound engineering principles instead of repackaging previous experience," Lloyd claims. It was also a policy that brought numerous brilliant and creative engineers and designers from around the world to their doors over the next 34 years.

3 DSMA - PHASE II

Candu reactors required continuous on-power refuelling. Project engineer Fred Hummel headed a team of AECL, Ontario Hydro and DSMA engineers to develop a robotic unit that could be remotely positioned in front of any of the 306 tubes to remove a spent fuel bundle. A second unit, also remotely operated on a rail, could inject a new bundle using the 225 inch telescoping snout.

While George V. Meagher (pronounced Marr) was the third and last person to become a partner in 1957, he was to emerge as the dominant individual of the partnership ending up as president and CEO of Dilworth Secord Meagher and Associates Limited, 1978, and chairman and CEO, 1987 to 1992. A fifth generation native of Halifax where his family had been merchants for years, George graduated from Dalhousie University and continued his education at McGill University where he obtained a degree in mechanical engineering in 1942. After a brief stint with Imperial Oil, he became assistant head of the engine laboratory of the National Research Council in Ottawa. An older sister Blanche, an M.A. graduate in political science from Dalhousie, started working that year in the Department of External Affairs and, as a result, George met many future members of Canada's diplomatic corps which was later to prove beneficial to the company.

George left the NRC to begin a varied career, including Canadian General Electric in Montreal, first as a division engineer and then manager of marketing research of one of their appliance departments. In 1957, during the construction of the trisonic tunnel at Ottawa, he and Paul Dilworth, who had become friends while they both worked at NRC, discussed the possibility of his joining the company.

"I had perceived the need for someone who could effectively perform both sales and public relations to get our name more broadly known, and to do an effective job at selling our services," Paul recalls, and it was agreed that George would become a third and equal member of the firm. Using the surnames of the three, a new legal agreement was drawn up by Paul's fraternity brother and long time friend Daniel A. Lang of the legal firm Lang Michener. Dilworth, Secord, Meagher and Associates (DSMA) was in business to take on any and all sorts of new projects.

One of the first new ventures was nuclear energy which "we felt we should get into if we could," Lloyd remembers. They agreed they should be primarily in the mechanical rather than the nuclear side and focused their attention on Atomic Energy of Canada (AECL) projects at Chalk River, Ontario, and Whiteshell Laboratories in Manitoba. They were also aware that Canadian General Electric in Peterborough had designed a prototype of the Candu reactor concept but not as an energy producing reactor, and that AECL had formed a new engineering division named Power Products.

"It," according to George, "had a mandate to design a nuclear power plant fuelled with natural uranium to cater to countries such as Canada. AECL eventually sold designs or turnkey operations to Ontario Hydro, Hydro Quebec, New Brunswick Power and overseas to India, in which DSMA participated." The company's involvement in the nuclear field began in 1958 when George Meagher saw John Foster, head of a joint AECL – Hydro program for a 200 MW nuclear station at Douglas Point and arranged to have Fred Hummel join the design team, with salary paid by DSMA, to work on the development of a fuelling machine. The work was carried out at AECL's North Queensway engineering office adjacent to Hydro's Kipling Avenue Research Centre, called Dobson Laboratories at that time. Fred, a 1939 B.Sc. graduate of the University of Saskatchewan, had worked for several firms designing industrial equipment and was considered an expert on mechanisms. "It was agreed he would be the key man in the design effort on the fuelling machine, while other work on the peripherals was carried out in DSMA's own office and laboratory, under Lloyd's direction," records Paul Dilworth. This nuclear experience later played a leading role in DSMA's participation in the United States space shuttle program.

Development of the fuelling machine for the Candu reactor was a highly complex under-taking that took some five years to complete.

The two teams realized that the Canadian reactor using natural uranium fuel moderated by heavy water, required exchanging the cylindrical spent bundles for new ones on a daily basis without shutting down the reactor. It was, therefore, obvious that a remote system was required, so the Hummel and Secord teams and others at the AECL and DSMA labs developed, over a five-year period, a fuel transfer system that could automatically make the selective removal and replacement of fuel bundles to the 306 tubes in the reactor.

Lloyd Secord summed it up more succinctly: "Canada originated this innovative system of remotely transporting fuel into and out of the Candu reactors that made it an economical producer of electricity, and it was this example of robotics operating in a hazardous environment that led to Canada's involvement in the Canadarm."

At about the time the Douglas Point reactor became operational, DSMA was embarking on a new contract with AECL to develop shaft seals for primary coolant pumps for nuclear reactors. It was Ian Billington's innovative concept to use a controlled leakage seal to recirculate minuscule leakage of heavy water in the pumps. He selected Rodney Bell to work with him. Later it became Rodney's major interest. An honours graduate of Oxford University, Rodney came to Canada when he became disheartened with his job at Shell Oil in Great Britain and asked Paul Dilworth's sister who was a friend of his father, if she knew anyone in the engineering business in Canada. She gave him Paul's home address and he got a reply on DSMA letterhead. Paul offered him a chance to come to Canada and work for three months to see if he liked it and that if he didn't, he would have enough time to find another job. Bell soon decided to stay after an initial project required him to study the design of a gas turbine compressor. He found this and other work mentally challenging and much more rewarding than his work in the U.K. "It was a fascinating company," he recalled. "They were doing stuff that was quite advanced and it really excited me."

His early assignments included the study of a bearing failure for a large paper machine and the development of a sound reduction system for a U.S. army gas turbine electrical generator. Rodney describes that machine as one with a six-inch diameter with a speed of 35,000 rpm. "Structural problems, such as disintegration during testing, caused an end to that project," he adds with a hearty laugh. He also took part in the aerodynamic design of the 9m x 9m low speed wind tunnel in Ottawa.

After about two years with the company Rodney learned FORTRAN, the computer software at the University of Toronto on an IBM 7094 computer. This was to lead to Rodney's major involvement in Ian's innovative work on shaft seals and hydraulic analysis at DSMA over the next eight years.

The first shaft seal contract was for primary coolant pumps used in nuclear stations. By 1965, the 20 MW nuclear power demonstrator plant was in operation on the Ottawa River and pump seals were a major problem. "It was too late to put the seals in the 200 MW Douglas Point as it was just starting up but the intent was to put them in Pickering," Rodney recalls.

The success of DSMA in nuclear projects prompted a series of nuclear contracts over the next 20 years. AECL, Canadian General Electric, Westinghouse, and the Hydro establishments of Ontario, Quebec, New Brunswick and Manitoba were among the organizations that called on

DSMA, and DSMA ATCON when it was formed, to do nuclear studies to meet a wide variety of needs. In 1970, for instance, the NUS Corporation of Washington, D.C., asked DSMA to be associate consultant for a study of Canadian fuel handling methods for the Japanese, then considering a 300 MW reactor. Fred Hummel accompanied a team of Japanese scientists to the Douglas Point, Pickering, Ontario, and Gentilly, Quebec, sites as well as attend a series of meetings with them in Washington.

At about the same time, AECL's Whiteshell Nuclear Establishment in Manitoba requested safety studies of the organic cooled reactor. This involved developing a test rig to simulate the rupture of a pressurized organic coolant pipe. Under the supervision of George Bremner, who joined the company in 1963 as an aerodynamicist, and would eventually become president of DSMA International, Ed Perchyshyn ran tests and Harry Goulding developed a computer model to simulate the test. Chalk River also requested analysis of the nuclear power demonstrator reactor when it was converted from a non-boiling to a boiling coolant reactor, with Rodney Bell converting the computer codes from the Gentilly reactor for use in this simulation. These programs started a long history of nuclear plant analysis at DSMA.

Rodney also became involved in the 250 MW Boiling Light Water reactor planned for Gentilly, when he was given the task of developing a 24 inch seal for the fuelling machine head. It also required a test rig to check out the seal design and analysis to overcome the large thermal distortions in the seal operation. These two developments also led to the design of a test rig for Bruce Point and other test rigs, as well as the design of a number of hydraulic seals and their analysis for various projects in Canada and offshore, Rodney recalls, adding that, "We were particularly interested in designing seals and selling them through Champlain Power Products

(CPP). CPP was the company that had been founded by Paul and Lloyd in 1954 to possibly provide ground power units for the *CF 105 Arrow.*

CPP had remained virtually dormant for a decade until reactivated in 1964, when Ray Hogue, a Toronto University graduate in mechanical engineering (1949), was hired by Paul Dilworth to be general manager of

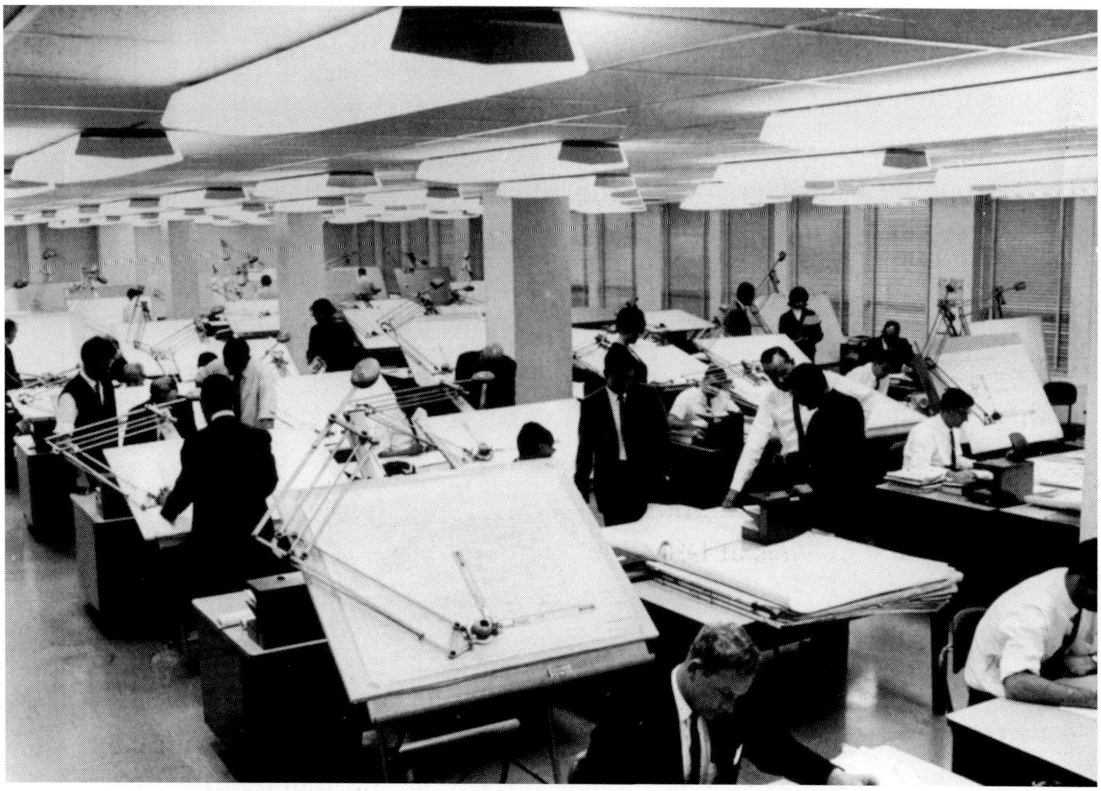

Drafting boards instead of computers dominated the office of Dilworth, Secord, Meagher and Associates at 4195 Dundas Street W. Today, only Roy Fedoruk keeps one in his office along with a computer. "It gives me something else to look at when I get tired of looking at the screen," he laughs. Photo:Courtesy W.A. Tony Zubrzycki

CPP after a career in the military, Sperry Gyroscope and Honeywell in Ottawa. According to a booklet produced shortly after Ray took over, CPP offered the "design and supply services on a wide range of products and systems in the fields of advanced mechanical engineering," and listed as its number one product, hydrostatic seals. These were the ones designed by Rodney and others but manufactured by CPP. Another product was the elliptical seal, also for nuclear reactors, "but also available for non-nuclear application" and the fluid velocity meter, which measures very low fluid velocities and is based on research work done at the National Research Council.

Rodney claims, however, that the installation of the hydrostatic seals in Canada was never very effective, mainly because the maintenance staff were not properly trained. On the other hand, those sold for use in the reactor at Lovisa, Finland, was a different story. "I was able to take some technicians and Cec Corke as manager of the labs over there and we trained the staff from scratch. They ran for more than a decade without a hitch," Rodney cheerfully remembers.

CPP, however, expanded to become an integral part of DSMA operations. When a contract was signed to design and engineer the blowdown wind tunnel for the Indian government at Bangalore in the early 1960s, CPP produced a number of precision controls required for it including a sensitive recording device to measure the temperature of the air during a test. It also won a number of direct contracts from AECL for various precision instruments and controls. In the latter 1960s, Champlain Power Products was restructured to become a supply and construction company capable of offering clients a complete package of everything from the construction of the building to the locks on the doors: in short to become a turnkey operation.

Designing gas dispersion systems to offset any potential escape of deadly hydrogen sulphide at heavy water plants was another DSMA development. Commissioned by AECL to come up with a solution, a DSMA team developed a series of propane burners that would automatically ignite the moment any leak of the deadly gas occurred, driving the hydrogen sulphide thousands of feet into the air. Two systems were actually installed, the first for Glace Bay, Nova Scotia, and a second at Port Hawkesbury, Ontario.

DSMA became involved in another major concern dealing with nuclear power: the hydrogen sulphide gas produced by heavy water plants was deadly if allowed to escape by accident or from the fractionating columns. The problem was first identified at the Glace Bay, N.S., heavy water plant where Rodney, in a preliminary study reported that if "the wind blew the right way, half the people of that city located half a mile from the plant could be wiped out." The plant was built by an American company who had believed they could use sea water to produce heavy water for the Candu reactors. However, salt destroyed the pipes and the plant was shut before it ever went into operation. Almost a decade elapsed before it was decided to refurbish the plant using fresh water and steps were taken to have DSMA make sure there was a foolproof system to disperse any possible leakage of hydrogen sulphide over the city itself.

DSMA put together a team for that project, initially under Ian Billington who conducted simulation tests in the DSMA laboratory. The project was later placed under the direction of Tony Stonehill, who conceived a system utilizing a ring of propane fuelled burners with automatic gas detectors which were erected around the plant. A small group of burners were tested during the winter of 1972/73 at a remote section of the Pearson Airport serving Toronto, and 101 burners were firmly installed around the plant in the summer of 1974. The matter of firmness was confirmed when a freak storm with winds of 115 mph an hour failed to cause any damage to them, and a full scale test using smoke instead of the deadly hydrogen sulphide was successfully carried out on February 15, 1975. "As a result, at the mere whiff of hydrogen sulphide the burners would automatically turn on and draw ground level air inside the ring, creating a flume to drive the gas some 10,000 to 15,000 feet skyward, depending on wind conditions where it would disperse harmlessly, " Lloyd recalls.

This led to AECL authorizing the development of two other gas dispersion systems to serve the heavy water plants at Port Hawkesbury, Ontario, and the proposed La Prade plant near Gentilly, Quebec. The latter, however, was never completed as the AECL plan to have heavy water available for foreign purchasers of CANDU reactors was scrapped.

While George Meagher was responsible for introducing the company to nuclear work, Paul and Lloyd also became extremely active in obtaining contracts in this field, which at one point, according to a company brochure, accounted for at least a quarter of the company's engineering activity. But Paul is quick to point out that, "George played a major role in getting us overseas business and brought us up from a company with no public image to one with a tremendous public image." George also encouraged the expansion of the company by forming subsidiaries or teaming up with other companies to form affiliates or partnerships.

By the late 1960s, the company was also deeply involved in another Canadian mega-project when Lloyd Secord negotiated a contract to be the prime consultants of a cyclotron with the acronym TRIUMF to be built on the University of British Columbia campus.

By 1974, the company, flush with a number of projects in the wind tunnel, nuclear and telescope fields, engaged an American management consulting firm to recommend a restructuring of the company. According to Uldis Kiesners, another of the design engineers to play a long term major role in DSMA, the company had grown "with Paul mostly involved in the company management, Lloyd looking after the nuclear work and George, the wind tunnel projects. All of them participated in business development," he added, "with Dr. Ian Billington the key person in the preparation of proposals and front end execution as far as aerodynamic design was concerned." Besides Uldis, the other two key people were Douglas D. Hunter, who joined the company in 1964 after 18 years with Dominion Bridge in Montreal, to be in charge of the mechanical design department and Jim Campbell heading the electrical controls group.

As a result of the management consultant study, the company was divided into two groups, Lloyd taking charge of the Advanced Engineering Group with a staff of 35, and George the facilities group with 13 people. The Engineering group was further broken down with Ian in charge of technology and Warren Brown heading a commercial unit, while George's small unit had George Bremner as manager of a facilities division, Uldis in charge of engineering and Doug Hunter heading the construction unit. "Paul Dilworth remained chairman of the board but was not associated directly with either group," Uldis wrote, and added, "This was essentially the beginning of the break up of DSMA into two separate companies." That happened four years later when Paul Dilworth retired.

In those four years, George went alone or led delegations around the world to bid on any number of possible projects. Both he and Lloyd made separate trips to Saudi Arabia when they heard its government was considering building a telescope which ended up in a small project more than a decade later. George also made trips to a number of other countries, among them Germany, India, Korea and Romania where he signed a contract in 1972 to build a blowdown wind tunnel in Bucharest. In 1978, George signed a contract for a blowdown wind tunnel in Yugoslavia, the first major wind tunnel contract signed by George as president of DSMA International, which had been formed earlier that year because Paul retired and Dilworth, Secord, Meagher & Associates was split into two: DSMA ATCON with Lloyd as president and DSMA International headed by George. This split occurred partly because George and Lloyd had different points of view on where they should concentrate their efforts for future growth. George charged ahead looking for overseas contracts, primarily in the wind tunnel business, and Lloyd concentrated his efforts on a broader range of projects including telescopes, space, and nuclear projects.

With the split, Ian Billington joined ATCON as a vice president and head of the company's laboratories called ATCON Services Limited while George Bremner became an Associate of the International group. The rest of the staff, which then numbered roughly 75 people, were either divided on the basis of their expertise, or experience or the project they were currently involved in, while others were allowed to elect their preference. Tony Zubrzycki, for instance, remembers being taken for a long lunch by Ernie Eggmann to persuade him to stay with the ACTON group and then getting into Ernie's Volkswagen beetle "for a trip home instead of back to the office."

4 TRIUMF and Partnerships

More than 100 people who were employed in building TRIUMF on the UBC campus in Vancouver took time out for a photograph, partly to show the enormity of the project. It required 4300 tons of steel. Opened in 1974, it has made not only astrophysical discoveries but has also contributed to medical and industrial research.

Another triumph of Dilworth, Secord, Meagher and Associates was TRIUMF– Canada's first cyclotron that is still proudly achieving scientific accomplishments 25 years after being officially opened on the University of British Columbia (UBC) campus. An acronym for the Tri-University Meson Facility, it was initially sponsored by three universities in Western Canada: UBC, Simon Fraser and Victoria, with the University of Alberta later joining the project. The Federal government, through its agency, the Atomic Energy Control Board, put up more than $30 million to build the cyclotron, with another $4.4 million contributed by the four universities for housing what, on completion, was the third in the world, and the world's largest.

Dilworth, Secord, Meagher and Associates, retained as prime consultants for TRIUMF opened an office in Vancouver headed by Fred Pierce, a former president of the Association of Professional Engineers of British Columbia. A staff of some 40 people was eventually involved in overseeing the on–site construction at UBC. Heading the on–site team was Tony Stonehill, while others who came from Toronto included Varis Smiltniecks and Walter Grundmann, an analytical engineer who joined the Dominon Astrophysical Observatory following the completion of the cyclotron.

The project was started in 1966 when the idea of using a major breakthrough in accelerator technology– the use of mesons for nuclear physics, nuclear chemistry and radiotherapy– was proposed by a group of UBC scientists. They made a preliminary sketch and model of what was required and approached DSMA to discuss the engineering aspects of the project. This led to a seven–year involvement to provide the design of the large cyclotron magnet, its support structure, and the support structure for the vacuum tank, under the direction of Uldis Kiesners who served as assistant project manager at the Toronto office. The Vancouver team worked on the design of the vacuum chamber, the magnet coils, resonators and controls.

It was a huge undertaking involving investigation of material availability, effect of chemical composition and microstructure on magnetic properties, Canadian manufacturing capabilities, and transportation and handling capacities. It even involved a study of earthquake considerations with readings taken at El Centro, California, and built into the design a provision to make the cyclotron earthquake proof.

At the Engineering Institute of Canada, in a presentation that won the Angus Medal as the best mechanical engineering paper of 1972, Uldis, said, "To achieve the expected performance of the cyclotron, a large magnet of approximately 850 inches outside diameter, 675 inches pole diameter and weighing over 4,000 tons is required." He described how these requirements could be satisfied most economically using three- and five- inch rolled low carbon steel plates. These were bolted or welded into blocks for construction of the six magnetic sectors that were then tied together at the magnet centre, each of them supported on one centre and three outer flexible supports to cope with thermal changes and earthquake conditions.

Contracts for the precisely rolled three inch plates, roughly 60% of the requirement, were awarded to Davie Shipbuilding of Quebec, while the thicker five-inch plates were ordered from Lukens Steel Company of Coatsville, Pennsylvania. "When the plates were bolted or welded together into blocks ranging in thickness from 50 to 80 inches," Uldis observed, "shimming between the plates was required in only a few blocks to achieve the specified tolerance which was 1/16 of an inch."

In 1970, Keith Lacey, a 1954 graduate in electrical engineering from the University of British Columbia, joined DSMA and returned to B.C. to design the control system and eventually become project manager. "It was the first cyclotron to be controlled by computers and I was hired because I had been working on computers at the Communications Research Centre of the Department of Communications in Ottawa and I wanted to get out of the government at the time." Over the next 15 years Keith and Jeff Stringer, a mechanical engineer trained in England who joined DSMA in 1967 and worked on the mechanical and structural requirements of TRIUMF at the Toronto office, were to become involved in several more accelerator and fusion projects undertaken later by ATCON.

In May 1980, for instance, ATCON announced its involvement in studies for the Canadian High Energy Electron Ring (CHEER) to be built at the Fermi National Accelerator Laboratory near Chicago. The Canadian contribution to a joint Canada/US project, its main component would be a closed accelerator loop, nearly one kilometre in circumference, around which a beam of electrons would travel at very high velocity. This ring would be tangent to a much larger existing American ring around which a high energy proton beam would travel in the opposite direction. The two beams would be made to collide head-on and the resulting particle reactions would be studied. The costs, however, were high for a relatively small community of scientists, so the project got cancelled.

Another project was the Electronic Ring of Saskatchewan (EROS), designed and built for the University of Saskatchewan. It was an upgrade for their accelerator to expand its capabilities and energy at target impingement. Doug Ferguson was moved from Toronto to Saskatoon to serve as project manager.

While not involved in the Eros projects, Jeff Stringer played a major role in a team of four to take part in a project known as the PPPL (Princeton

Plasma Physics Laboratory), for Princeton University between 1983 and 1985. There was a need for a robotic device to replace ceramic tiles which lined the interior of their TOKAMAK reactor. The fusion reaction releases very high energy neutrons which causes the lining and much of the TOKAMAK to become highly radioactive. "The device for removing the tiles, therefore, had to be robotic," Jeff recalls. "There were four of us on the job the first year, myself, two from Spar Aerospace Limited and another chap from CAE in Montreal, and three of us in the second year. We came up with a maintenance manipulator, somewhat like the Canadarm, to remove the tiles."

Meanwhile, at home, the University of Alberta proposed a major project called "MARIA," that began in 1981. The title stood for "Medical Accelerator Research Institute in Alberta" and the plan called for a pair of cyclotrons to promote "advanced research in medicine, radiopharmacy as well as to be a specialist centre to evaluate new methods of diagnosis and therapy and to care for patients who have malignant diseases." ATCON and its manager in Edmonton, Keith Lacey, were deeply involved providing a conceptual design of the project that was being promoted by the Provincial Cancer Hospitals Board and the University of Alberta.

Two groups were to use it: the medical people during the day and physicists at night. "The medical people didn't want the machine to shut down for more than about 24 hours which meant you couldn't let it cool down radioactively," Keith remembers. "That meant you had to do hot changes and you had to have duplicate critical parts so the cost was high." Despite that, by early December 1981, a slick eight-page brochure was published highlighting all the reasons why it should be approved. "But before the month was out, Premier Peter Lougheed called us in and said that he was going to have a budget shortfall the following year so the project was a no go," Keith adds.

The next proposal that involved Jeff arose when a KAON Factory, an acronym for kaons, antiprotons, other hadrons and neutrinos, was proposed for the UBC campus as an extension or expansion of TRIUMF. The word factory was added because it would "produce particles in unprecedented quantities."In 1985, the Federal and British Columbia governments agreed to share the cost of an $11 million engineering design and impact study, and Jeff became one of the people involved as an employee of ATCON. A leaflet promoting the project claimed it "would join a large network of large new accelerators proposed around the world" and suggested building two new "synchotron" accelerators "to increase the energy of protons from the TRIUMF cyclotron sixtyfold, from 500MeV to 30,000 MeV (30GeV)." While Jeff was still working on the study at the KAON office in Vancouver, ATCON ceased operations early in 1986 but Jeff was retained at UBC to work until that Fall. He left when offered a job with Wardrop Engineering in Toronto, which, on the closing of ATCON, had taken on Lloyd Secord as a consultant and other ex–DSMA ATCON employees, Tony Stonehill, Tony Zybryzcki, Carmen Volpe, Cecil Corke, Ray Coladipietro and Peter Bolden. Meanwhile, studies continued on the KAON project for years. It was finally cancelled in February 1994 when the Federal government claimed foreign governments did not make sufficient financial commitments to keep alive what had become a $750 million project.

TRIUMF, however, continues to pave a major role in nuclear research. According to an article in the *BC Business* magazine, to mark the 25th anniversary of TRIUMF in 1999, the cyclotron has not only satisfied scientists pursuing pure science but has also contributed to a number of practical developments. In the pure science area, it has raised questions on such things as why quarks join to form more complex particles and has recently opened a $10 million lab called an Isotope Separator and Accelerator (ISAC), to measure "nuclear reactions relevant to astrophysics and the opportunity to produce large yields of rare short-lived radioactive isotopes."

In the medical field, it is given credit for treating brain tumours and several types of pelvic tumours and the development of a positron emission tomograph (PET) scanner gave researchers the capability to study brain disorders such as Alzheimer's and Parkinson's disease. According to the *BC Business* magazine article, its proton therapy facility used by the B.C. Cancer Agency successfully treated people across Canada suffering eye cancer. With respect to cancer, the scanner also annually produces many millions of dollars worth of isotopes for medical applications around the world. This means that instead of conventional treatment, which bombards the body from the outside with radiation and ravages the immune system, it produces "radioactive seeds" that are surgically inserted directly into the tumour.

In industry, TRIUMF in 1998 helped build a large radio frequency oven chamber for a farm in Peace River, Alberta "which now dries hay with a marked improvement in quality at competitive prices." Another development along similar lines for the forestry industry, "can dry wood more quickly, uniformly and efficiently, " the article reports.

TATA-Dilworth, Secord, Meagher and Associates

In 1974, the same year that TRIUMF was completed, one of two DSMA partnerships was severely jolted when India exploded a small atomic bomb on the edge of the Rajasthan desert. DSMA's association with India was more than ten years old by then as DSMA had designed the blowdown wind tunnel for India's National Aeronautical Laboratory at

Bangalore. By the time of its completion in 1968, George Meagher was involved in developing a consulting firm on the lines of DSMA in India, by teaming up with Tata Sons Limited, the biggest and most influential industrial complex in India with headquarters in Bombay. A seven-man executive committee of four Indian nationals and George Meagher, Senator Daniel Lang and Douglas D. Hunter for DSMA was established in May 1969 and, within a month, Doug Hunter arrived in Bombay as the Canadian Engineering Manager.

Various potential projects were identified by Doug Hunter. They included a revised fuel handling system and end shields for an atomic reactor at Rajasthan, some work on a variable energy cyclotron, two furnaces for determining the fire rating of building materials, and design assistance for Hindustan Aeronautics Limited on jet engines for a fighter aircraft. Other visitors to India, besides George Meagher, included: Fred Hummel to look into projects involving atomic energy, Dr. Ian Billington to discuss engineering assistance for a space centre at Thumba, and Doug Knowles to talk to senior air force officials about the jet engine program - but most of the proposals failed to materialize except those related to nuclear reactors.

On Saturday, May 18, 1974, however, all that activity was dealt a severe blow when India detonated a small bomb on the edge of the Rajasthan Desert. There had been rumors that India was attempting to make such a bomb, particularly since it had refused to sign a non–nuclear agreement. In fact, according to Prime Minister Pierre Trudeau in his *Memoirs,* he had earlier asked Prime Minister Indira Gandhi point blank: "Are you working on a bomb? She gave me her word that India was not doing anything like that." A short time later, in 1974, his *Memoirs* continues, "off went this thing that sounded an awful lot like the test of a nuclear bomb, but that the Indians called a 'peaceful' nuclear explosion. I guess that was how she was able to play with words and give me her assurance that there was no bomb in the making."

The Canadian government, however, in the midst of an election campaign, didn't play with words. By the following Thursday, External Affairs Minister Mitchell Sharp announced suspension of shipments worth $3 million dollars of nuclear-related equipment. He also called for a review of the $70 million loan program to aid India's transportation and communications, mining and manufacturing and electrical power centres.

The head of India's Atomic Energy Commission, H.N. Sethna admitted the blast was produced at a natural uranium research reactor at Trombay, near Bombay, that was built with Canadian help, but denied India had broken any international commitments. He claimed that the plutonium was produced on the second fuel cycle in the Trombay reactor while the agreement with Canada restricted the "use only" of plutonium produced on the first cycle. His position is supported by George Meagher. "Yes, there was a loophole in the earlier agreement," he says. Mitchell Sharp as Minister of External Affairs also agreed that the test did not violate any agreements with Canada, but was contrary to the peaceful purposes for which Canadian nuclear aid was intended.

As a result, financial and technical aid in developing atomic energy for India that had been a Canadian policy since 1956, was withdrawn. "So we had to withdraw our people and stop the flow of technical data," George recalls. "We tried to refocus on the aerospace market and obtained another wind tunnel contract, but then Joe Clark, as Minister of External Affairs, cancelled our permit in mid-project. We never fully recovered from these blows - indeed Canada has never fully recovered the preferred position it had with India,"George claims.

Pengalta Research and Development Limited

While the partnership Tata–DSMA was being brought to a virtual standstill in 1974, Dilworth Secord Meagher and Associates had entered another partnership with a large Canadian engineering consulting firm, Reid Crowther & Partners Ltd. of Calgary. Lloyd and Ed Crowther, who had joined Turbo Research at the same time in 1945, established an affiliated company, Pengalta Research and Development Ltd., with each partner providing two directors. Paul Dilworth and Ian Billington were the first two to represent DSMA interests, with Ian later becoming president. Most of the early engineering activity for Pengalta was subcontracted out to one of the two partners. The first objective of the new firm was to extract helium from natural gas pipelines in Alberta as there was a belief that helium from wells in Texas was running out.

The Alberta Research Council, with provincial government backing, built a laboratory model to demonstrate the effectiveness of stripping helium from natural gas flows by permeation through a thin plastic membrane. Pengalta was then selected to coordinate the building of a portable pilot extraction plant for Alberta Helium Ltd., and much of the design work was to be done by DSMA at Toronto.

From the decision to proceed with the project in 1974, it took almost six years before any attempt at helium extraction took place. By then the work at DSMA had been moved to ACTON with Peter Tillson one of the key engineers to develop the controls that were required to hook the system up at the MacLeod River Compressor station near Edson, Alberta. "It was complex because there was a demand for not only a manual and remote control system at the MacLeod River site, but also a remote control system to be activated at Edmonton, some 200 miles distant," Peter recalls. Helium from this pipeline source had concentrations of the order of one part in one thousand. With three stages of molecular filtration that included bypass recycling, the final concentration achieved 99% purity, making it uneconomic to compete with Texas which had found a way to extend its helium supplies. "They ran the plant for approximately a year before closing it," Peter recalls, "but at least they proved they could get helium out of it. And it was a very interesting project for me personally," he adds.

Before the decision was made to close that plant, another major Pengalta project was well underway: the development of pumps to extract heavy oil from the oil fields on the Alberta/Saskatchewan border surrounding the city of Lloydminster. This work got underway in September 1977, when Murphy Oil, with extensive oil properties in the area, asked Pengalta if they could solve the problem of pumping extremely viscous oil from their wells. More commonly described as heavy oil, the company had tried such things as injecting steam into the reservoirs to reduce the viscosity, but still found standard pumps unable to work effectively in bringing oil to the surface.

A company named Pengalta was formed by DSMA ATCON and another Calgary consulting firm, Reid, Crowther & Partners Ltd., to extract helium from natural gas. A complex series of pipes housed in a building near Edson, Alta., that could be controlled by computer on-site and at Edmonton, some 200km distant, were developed, with a major part of the control work being done by Peter Tillson of DSMA. Photo: Courtesy Peter Tillson.

Realizing that other oil companies had or might have the same problems, Ian Billington and Jim White, a University of Toronto graduate in Engineering Physics who had joined DSMA in the early 1970s, eventually got a number of oil companies to form a consortium to finance the project at Pengalta. A contract was awarded to ATCON and by January 1979, the company newsletter reported that various transparent models of oil pump components had been constructed by Ed Perchyshyn in the lab with "very interesting results obtained" that might contribute to the development of improved bottom hole pumps for lifting heavy oil.

A second phase project report in December 1980 indicated a new laboratory was under construction at the Foothills Industrial Estate in Calgary with the Pengalta head office being moved there as well. Four months later, another report indicated that not only the pump test rig designed by Ed Perchyshyn was being installed, but also one designed by Dan Simec of ATCON was also being assembled at the Pengalta laboratory. A project sheet reported that it was designed to simulate a down hole pump for depths of up to 2600 feet. However, support for the project slowly evaporated and according to Lloyd Secord, who replaced Paul Dilworth as a director at Pengalta, "some pumps were sold to firms pumping heavy oil in Venezuela and worked well." Several Alberta firms tried them as well but lost interest when they found heavy oil production couldn't compete with conventional crude oil that was falling in price at that time.

5 DSMA ATCON

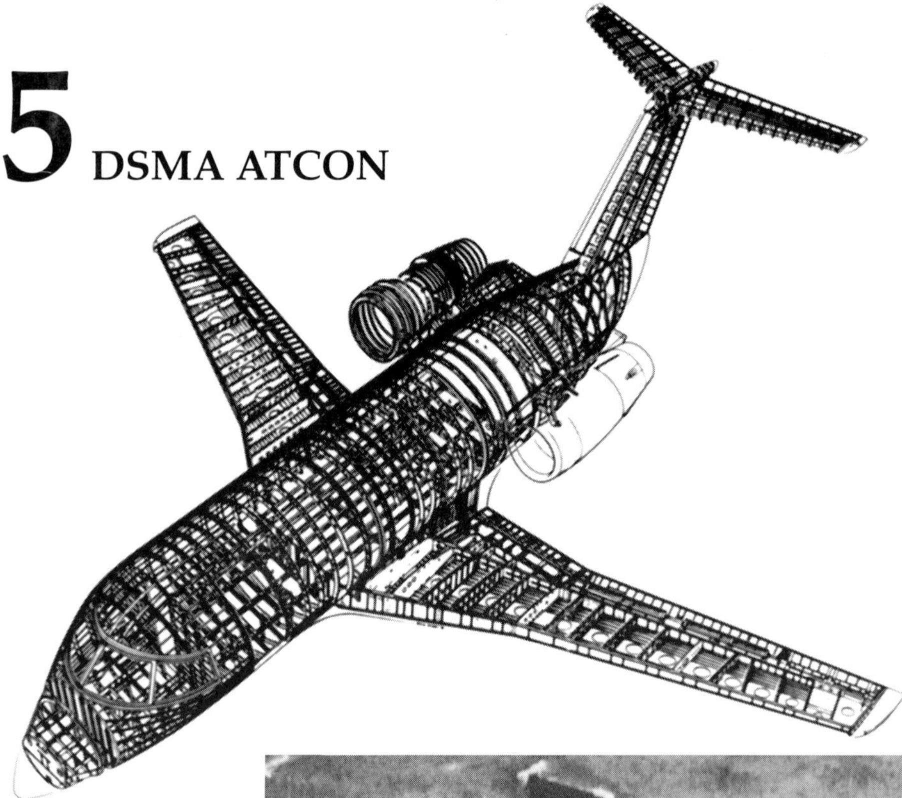

The superstructure of the Canadian Challenger, the world's first wide body business jet is shown above. DSMA ATCON wrote the final stress analysis reports for the fuselage, including the pressure shell (skin and stringers), floor and engine support structures, nosewheel well and associated frames, required before Canadair's submission to Transport Canada and the F.A.A. to obtain airworthiness certification. The photo shows the aircraft in flight. Photo: Courtesy Canadair.

When Paul Dilworth decided to retire, Dilworth, Secord, Meagher and Associates split into two entities, one of the two new firms being named DSMA ATCON LIMITED, the word ATCON being short for Advanced Technology Consultants. It inherited the nuclear projects awarded earlier by Atomic Energy of Canada Limited (AECL) and the Hydro Electric Power Commission of Ontario. Lloyd Secord was named president. A native of St. Thomas, Ontario, and the gold medal winner in mechanical engineering at Queen's University in 1945, Lloyd was considered by many colleagues as the "consummate consulting engineer," and over the years proved to be a successful entrepreneur when it came to seeking contracts in emerging technologies.

Besides the nuclear program, ATCON also inherited work being carried out on the Canadarm project which a decade earlier became Canada's entry into space robotics. This occurred because Lloyd got DSMA involved in astronomy and nuclear projects through securing the design of an automated manipulator for NASA's Large Space Telescope program that later became the Canadarm project. Working closely with SPAR Aerospace Limited were a number of ATCON staff designing not only the equipment to test the arm as if it were operating in zero gravity while still at ground level, but also designing other aspects of the arm itself, such as the end effector or "hand" and controls for the shoulder, elbow and wrist. It was another reason the company was seen as one of the finest engineering-consulting firms in Canada. As Harvey Griggs, present Chairman of H.G. Engineering and a key player in the establishment of the Aiolos company in 1994, recalled, "they raised the standard of mechanical engineering consulting to an entirely new level."

"Some of their stuff for the nuclear industry was absolutely wonderful," Harvey adds, citing their robotic refuelling system for the Candu nuclear reactor, the development of gas dispersion systems for Canada's heavy water plant sites, and dozens of other projects in the nuclear field as examples of creative and innovative engineering that was admired in Canada and abroad.

An illustration of Lloyd's innovative engineering skills followed in the wake of the Mississauga train derailment of November 1979 and the escape of chlorine gas from a tank car which forced the evacuation of thousand of residents. He authorized a study to design a portable gas dispersion system as a possible unit for fire departments. One such portable system was developed in ATCON's own laboratory, ATCON Services Limited, but it proved too costly to interest any private or government organization.

Nor was the work confined to hydrogen sulphide or chlorine gases being blown into the atmosphere. Canada's Atomic Energy Control Board authorized a study of radon gas emanating from landfill at Elliot Lake, Ontario, another for the Eldorado Refining Company in Port Hope, Ontario, and a third at Uranium City, Saskatchewan, where the mining of uranium was taking place. These projects, promoted and contracted jointly with Acres Engineering of Toronto, also led to the creation of a subsidiary company in the United States in 1979, when the Environmental Protection Agency of that country asked ATCON to assess radon gases emanating from the phosphate mining area of Polk County, Florida, between Orlando and Tampa. The US company was called American Atcon Inc., with the work there led by Arthur Scott, who later bought the company. Gases were identified at all three locations and the problem was resolved with the enactment of improved building codes such as radon resistant foundations and changes in building ventilation systems, based on solutions learned in remedial work at Elliot Lake and Uranium City.

In the early 1980s, other studies were commissioned to assess methane gas at an Etobicoke construction site and another 30-acre area in downtown Toronto where methane problems were identified. They were the result of "the past use of a variety of fill and the remains of now-vanished industrial installations," the ATCON newsletter reported. The use of urea formaldehyde foam installation (UFFI) in homes, which became a contentious issue, was also an ATCON study in the early 1980s. According to the same newsletter of April

1983, ATCON and Acres Engineering "began surveying new homeowners who requested it late in 1982, and checked about 50 homes a week," and added, "Where formaldehyde installation was suspected, they drill about a dozen holes to verify the presence of UFFI."

Dozens of other projects originating at DSMA became the recognized expertise of the successor, DSMA ATCON LIMITED. Between 1973 and 1975, DSMA was asked to carry out assignments for Eldorado Nuclear Limited to design casks to ship spent fuels from nuclear power stations to reprocessing plants. DSMA was also commissioned by Ontario Hydro to design, manufacture and commission an on-site spent fuel transfer flask. By January 1976, photographs were taken of dropping the proposed flasks from a height of 26 feet onto reinforced concrete with a one-inch thick steel plate grouted into place. The flasks didn't break. Later that year, other projects in the nuclear field were awarded, one to provide an environmental assessment of the New Brunswick Nuclear Power Plant at Point Lepreau. Mathematical models were used to predict the rise in water temperature downstream from the plant under ambient conditions, taking into account the river's topography, tides, winds and ice conditions. In the same year, Eldorado Nuclear Limited awarded a contract to design modifications to an existing storage area to provide increased physical security for enriched uranium to be used in the production of Candu booster fuel.

As mentioned, when the split to become two companies was made, the nuclear expertise stayed with DSMA ATCON LIMITED which continued to carry out numerous contracts in that field. When it was discovered, for instance, that the fuelling tubes in the Bruce and Pickering reactors were coming off their bearings because a chemical reaction had caused them to stretch and begin to sag, ATCON and its laboratory, ATCON Services Limited, were given the job with two other consulting companies of re-shifting them back onto their bearings. ATCON was assigned the designing and building of the multiple head shifting machine for Pickering and to design and demonstrate a manual tube shifting operation for Bruce, the latter called the Bruce West Shift program.

"That made it two problems," recalls Al Bullock, who was made project manager of the job. A 1953 B.Sc. graduate of the University of Toronto who joined DSMA in 1965, he is still active with Aiolos. "At Pickering, there was a need to be able to undo the bolts, shift the tubes, then rebolt them again. At Bruce, it was a matter of developing equipment to slice the weld, hold the tube in place and reweld it." Another present-day employee of Aiolos, Roy Fedoruk, the design engineer for the Bruce project, who had joined DSMA in 1964 on graduating from Ryerson Polytechnical Institute, played a leading role in designing the required tooling to do the job.

At Bruce, two shielding cabinets made of heavy steel plate to protect workers from radiation were made and mounted on the fuelling machine bridge. The front wall of the cabinet was fitted with leaded glass viewing ports and rotating disk and turret to permit access to the fuelling tube. A machine was also developed to slide several feet along the tube to cut the weld. Another machine called an X-Y positioner, designed by Hydro engineers and made at ATCON Services, then moved and held the tube in place while two operators used a buffing machine to prepare the area for re-welding and then used a specially designed welding machine to complete the job. Also employed on this project was a former DSMA employee, Jack Richman, then working for Spectrum Engineering in Peterborough, who developed a robotic system that could be adjusted to go down the tubes several feet, cut the weld, do the shifting and reweld the tubes into place. After extensive testing at Ontario Hydro's Shorncliffe laboratories, the robotic machine was selected as it was considered the safest in terms of radiation exposure.

Al recalls that nothing was left to chance as the project took several years of conceptual and detailed design drawings. Two wooden mock-up cabinets were erected in the ATCON Services, complete in every respect to the units designated for use in the reactor vault. Hydro employees who would be doing the work then came to the lab to become familiar with the precise routine they must follow before the work was carried out in the early 1980s.

For the Pickering site where the tubes were secured by bolts, ATCON designed and built, under Doug Knowles' direction, a special purpose machine that could manually reposition all the fuel channels in the four reactors. Called a single shifting tool, the complex machine permitted workers to latch onto the fuel channel end fittings. It also had dual drive air and stepping motors to enable the workers to quickly connect onto or disconnect from the fuel channels once the tubes were shifted and once again bolted securely in place.

Two other unique machines to transport nuclear waste were also challenges for ATCON personnel in the early 1980s. AECL asked the company to "design a flask transporter and erector" to move cobalt (irradiated by being placed in a flask and inserted into the Candu reactor) from the reactor buildings to the spent fuel storage bay in another building. Roy Fedoruk was again assigned the job of project manager. With a team of engineers and designers, a transporter was developed that could receive the cobalt flask, measuring 27 inches in diameter and 19 feet long, in a vertical position and lower it to horizontal position for transporting to another building. There, the flask, which weighed roughly 25 tons, had to be raised once more to a vertical position so that the cobalt could be winched out of it by the flask's own equipment.

Nuclear problems, however, did arise over which the company had no control. One happened in 1979, when a melt down occurred in the reactor at Three Mile Island near Harrisburg, Pennsylvania. Overheating caused by an interruption in the reactor's cooling system caused severe damage to the reactor core, and while the protective building housing the reactor largely prevented any radioactive debris being released, newspaper headlines and TV and radio commentaries across North America spawned protest groups and nay-sayers by the thousands. As a result, for several years the expansion of nuclear generated electricity, worldwide, was put on hold.

Despite this and related setbacks, ATCON managed to retain staff and keep going because of its huge involvement in the Canadarm project, continuing work in fusion studies, the designing of telescopes and a number of smaller projects.

One of the latter, started in 1979, was a four– to five-year program to promote the development of electric vehicles sponsored by Transport Canada Research and Development Centre in Montreal. Marathon Electric Vehicles Incorporated, of St. Leonard, Quebec, supplied a number of its state-of-the-art C360 vans to the St. Lawrence Seaway Authority at St. Catharines, Ontario, to be used to transport men and materials during servicing operations on the locks and lifts bridges of the Welland Canal. ATCON was commissioned to develop instruments to automatically record performance data for the initial tests. Marathon would then develop further improvements and a second series of vehicle tests would be evaluated by ATCON so Marathon could develop a prototype of the final vehicle. A number of people became involved: Peter Tillson and Carmen Volpe at the ATCON office and Cec Corke and Colin Martin at the Atcon Services laboratory, among them.

In 1980, Tillson also became deeply involved in designing and installing the controls, instrumentation and acquisition system for a thermal vacuum chamber to be located in the Communications Research Centre at Shirley Bay, near Ottawa. Needed to test a new generation of communications satellites, the huge 22 X 35 foot chamber was being built by the High Vacuum Equipment Corporation of Hingham, Massachusetts.

In the late 1970s and early '80s, headlines were being made by employees concerned about working with video display units (VDTs). ATCON also became interested in this occupational environmental concern and carried out an assessment of the use of VDTs at the Toronto newspaper, *The Globe and Mail,* in the summer of 1980. ATCON's own newsletter reported, "This survey of *The Globe and Mail*'s VDTs for radio frequency and X-Ray emissions showed that all were performing very well with truly negligible emissions." The newsletter added that it had also checked its own office equipment and wrote, "It may be reassuring to know that a comprehensive survey, including our own computer, word processor, and a couple of home TVs, microwave oven and such, shows a no-problem result." It also tested a couple of Walkie-Talkies and CB radios which, it stated, "showed measurable RF (as they should because their purpose is to broadcast on radio frequencies) but they too were within the recommended guidelines.

There were numerous other studies carried out by ATCON personnel throughout the early 1980s, among them solar energy for Canadian Mortgage and Housing Corporation headed by Jim White who eventually left ATCON to join that organization. There was also a venture that enabled the company to participate in a solar aircraft proposal and stress analysis work was another area that ATCON carried on during the economic turndown of the early 1980s.

6 The Canadarm

Besides participating in the design and construction of the first Canadarm with others at Spar Aerospace Limited, DSMA ATCON designed and built the five rigs required to test the Canadarm at zero-gravity on earth. The design of the test rigs won the 1980 Award of Excellence from the Association of Canadian Consulting Engineers of Canada. Photo: Courtesy Spar Aerospace Limited.

Much has been written about Canadarm, the remote manipulator system developed for use in NASA's shuttle program that was built by Canada's Spar Aerospace Limited and worked flawlessly when tested on the second flight of the shuttle Columbia. Spar Aerospace received numerous headlines and accolades for its success, but had it not been for Lloyd Secord's initiative the Canadarm project might never have come to Canada.

It began when NASA's chief administrator visited Ottawa in 1969, as the space agency was looking for international involvement in the post-Apollo program and Lloyd, with telescope experience and aware that NASA wished to launch a large space telescope as part of that program, arranged through the Canadian government to visit the Goddard Space Flight Center in Greenbelt, Maryland. There he met with Frank Ceppolina, head of the team planning the servicing and refurbishing of that project. Lloyd went intent on getting DSMA a contract to participate in the design of the telescope. When asked what experience DSMA had, he showed photographs of the telescopes they had worked on as well as the robotic system his company had developed with AECL to refuel the reactor at Douglas Point, Ontario. Ceppolina was intrigued when he saw the robotic handling equipment designed for the CANDU reactor. He said he had seen nothing like it in the United States, and suggested Lloyd forget about designing telescope components as "we've got too many people around here with expertise in space telescopes."

He suggested, instead, that Lloyd get the Canadian government to support work on a robotic device to refurbish the space telescope from the shuttle's cargo bay as it was going to be too costly and time consuming to have astronauts working on them in space. When Lloyd pointed out that DSMA's experience was in the aeronautical field rather than space expertise,

The engineering model of a special purpose manipulator to service the Hubble Telescope for NASA in 1972 was a far cry from the eventual development of Canadarm that was successfully tested in 1981. DSMA and DSMA ATCON developed the original version of this manipulator, shown here in a model of the shuttle developed by the North American Rockwell Corporation in California. Photo: Courtesy North American Rockwell Corporation.

Frank pointed out that was a plus as the shuttle was intended to operate more like an aircraft than a costly space vehicle. "He also said they would want a package deal so we should bring in a hardware manufacturer to build whatever we designed."

Lloyd returned to Canada and initially talked to Dick Richmond, head of the Canadian division of Douglas Aircraft (later president of Spar Aerospace). They were interested but later turned it down as it didn't fit well enough within the Canadian company's scope of activities. Lloyd had also contacted the Department of Industry Trade and Commerce who agreed to authorize a small contract of $15,000 for DSMA to do some preliminary design work, thereby making DSMA, in the eyes of NASA, a legitimate government contractor. Lloyd took the design back to Ceppolina who agreed that "it looked pretty good" and suggested DSMA make a mock up of the design and "we'll see how it fits in the storage bay of the shuttle prototype being built by Rockwell in California."

To replace his Douglas Aircraft approach which, by now, had become McDonnell Douglas Corporation, Paul called Dr. Philip Lapp, a director of Spar Aerospace to begin what became a long term relationship as John D. MacNaughton, vice-president of marketing and planning of Spar, welcomed the opportunity to become involved. Although the Goddard team was not demanding space experience, Spar's space background and business acumen in securing significant government support turned out to be a good choice.

MacNaughton was a former member of the management team of the Special Products and Applied Research department (Spar) of de Havilland Aircraft Ltd. When, in 1967, de Havilland decided to sell that division to Larry Clarke, one of its senior vice-presidents, Spar Aerospace Ltd. was born with MacNaughton taking on the marketing and planning function. With the decision that Spar would participate, its president, Dick Richmond, who had left Douglas Aircraft, felt that John, as a vice-president of marketing and planning and the active chairman of the Air

Industries Association of Canada, should play a leading role in negotiations with Canadian government officials and NASA, Paul Dilworth recalls, "John's skills in leading the selling operation was executed with consummate skill."

"The initial design had nothing to do with the Canadarm as we know it today," Lloyd admits. Nor did it resemble an arm. It was a rectangular framework hinged at the bottom so it could be rotated up from its restraints when the cargo bay doors were closed during launch and re-entry. The manipulator would move along orthogonal axes to remove and replace scientific instruments or power packs. The original concept was for a unit that would be in the cargo bay of the shuttle only when the mission called for telescope servicing. When that concept was put aside, interest switched to a design that could function like an arm permanently attached to the shuttle itself.

For a while, it looked as if Lloyd or John wouldn't have to worry about having any arm when NASA decided to realign responsibilities at its various space centres and moved the telescope project to its Marshall Space Flight Center in Huntsville, Alabama. As this was the centre where the astronauts trained, its director made it clear he wasn't interested in having a robotic unit around his telescope because he felt his astronauts were better suited to doing the job.

"But," Lloyd recalls, "NASA's International office stepped in, disturbed that they were losing Canada as the only country to join in what was intended to be a departure from their wholly national *Apollo* program and be international in scope and funding. They also knew of the space competence of the Canadians who had played a key role in the launching of the *Apollo* project after the cancellation of the Avro *Arrow* in 1958." NASA got our participation back on track," says Lloyd, "by offering us the opportunity to replace the telescope manipulator with the robotic arm then described as the Shuttle Remote Manipulator System. This was a much bigger undertaking for Canada, both technically and financially," Lloyd adds, "as it was a critical component of every shuttle launch." Lloyd also remembers there was strong lobbying in the United States that the technology should be developed there but "Canada's reputation for successfully completing space projects, primarily in the communications field, plus NASA's wish to avoid losing its international identity won out."

In 1974, it was decided to add two more companies to the team - RCA Canada of Montreal, then actively involved in communications satellites, and CAE, a world pioneer in flight simulation also located in Montreal. "Since we were aware that the government would not want to contract with a partnership team of four, it was then agreed at a team meeting in the RCA board room that Spar serve as the major contractor for the team," Lloyd recalls. It was the same year that Canada's participation in the project got a further boost when the Honourable Jeanne Sauve, as Minister of Science and Technology, signed a space policy paper and made available $1,000,000 for studies on the shuttle manipulator program. This led to a memorandum of understanding signed between the Canadian government and NASA in which Ottawa's National Research Council was assigned to manage the project.

As the first phase of the project got underway, the U.S. head office of RCA decided to sell its Canadian branch, whose forte was in communication satellites, and Spar acquired the firm, shrinking the team to three with Spar to prepare the proposal to submit to the Federal government. "We opted for a major role in mechanical design and left it to Spar to decide how best to apportion the other major aspects including controls, manufacture and testing between themselves, CAE and DSMA," Paul Dilworth recalls. "But to our considerable chagrin, and that of CAE, when negotiations with Ottawa were consummated, Spar emerged as prime contractor and the other two members were relegated to and treated more as sub-contractors than team partners throughout the ensuing program."

The ensuing program was a long one for it took another seven years of design, manufacture, testing and qualification before the first manipulator was shipped to Florida, attached to the Columbia for its second launch into space, and performed flawlessly under its new name, Canadarm. Over that period more than 25 DSMA ATCON employees, engineers, designers, technicians and support staff under Lloyd's overall direction were involved in building and testing five major test rigs or were seconded to Spar's own plant to tackle various thorny issues. The centrepiece of the test equipment which DSMA designed and built was the System Test Rig to simulate zero gravity that the arm would be required to function in when operating from the shuttle.

Despite such a major involvement by DSMA and ATCON, and a memo requesting that its role be recognized in a Spar press kit prior to the first test of the arm, the role of ATCON and CAE was only mentioned on pages 13 and 14 of the 15-page kit. Spar, on the other hand, basked in the success of the project that became an ongoing success story as a contract was eventually signed to make four more Canadarms and carry out ongoing maintenance of the units. Spar's successor, MacDonald Dettwiler Company at 9445 Airport Road, Mississauga, has in fact recently refurbished the first successful Canadarm for NASA by increasing its payload capacity and updating its electrical systems for what is hoped to be another 20 years of continuing flights into space.

The 50ft., 9in. Canadarm is shown installed along the left wall of the shuttle Columbia *prior to its first successful test in November 1981. Photo: Courtesy NASA*

It is also a fact that Spar's founder and chairman, Larry Clarke, now retired, ended up anxious about being prime contractor. Wanting MacNaughton back in his role as vice-president of marketing and planning, he is quoted in a book published in 1992 to mark the 25-year history of Spar, as telling John to "get the project on track and work his way out of it as soon as possible." That's because Spar had been founded to develop expertise in satellite communication projects, at which it had become immensely successful, and was worried involvement in something as remote as the robotic arm might tag the company with a loser.

John confirms that the Shuttle Remote Manipulator System "was in a sense a bit of an orphan: it was not the mainstream of the company. I think there were quite a number of people in the company who thought

W.A. " Tony" Zubrzycki demonstrates how, while playing with an elastic band, the idea struck him as a possible device to develop the hand or end effector (below) for the Canadarm that was used when the arm was tested for the first time in November 1981. Photos: Courtesy Mel James and Spar Engineering.

it was dumb, or that it would never come to anything." He and Spar's president, Dick Richmond, however, worked diligently with the designers and engineers of the three companies over the next seven years before the first successful test flight took place. Every month senior management of the three companies met to discuss project schedules and review any problems that arose. "When a slippage or other problems were identified," Lloyd recalls, "it sometimes meant providing engineers and designers from our firm to work with Spar's people. At one time, I believe we had 13 of our staff working at the Spar buildings at Caledonia Road or Ormont Road sites."

Keith Lacey was one of them. He spent more than a year at Spar, and at one time he was manager of the Canadarm Interface Control group. " We had the job of matching and coordinating the physical, electrical and thermal interfaces between the arm and the shuttle with the various design groups involved," Keith recalls. "Sometimes, when a particular problem arose, and needed to be solved in a hurry, like a week or two, it was my job to form what was known as a 'tiger team' which meant –call whatever company had the best expert to work on the problem and request their participation. Sometimes I would be refused by the person's immediate supervisor and then I'd suggest that someone will be in touch with your president. That invariably worked," Keith laughs.

The team approach, with expert designers, engineers, draftsmen, machinists or whoever or whatever was needed, was a common occurrence during the construction of the arm. In the summer of 1975, W. Anthony " Tony" Zubrzycki, for instance, who had joined DSMA as a designer in 1966, was asked by his supervisor, Douglas W. Knowles, to report to Spar the following Monday to work on the " hand" or what was called the " end effector" of Canadarm. He was asked because he had previously contributed to the design studies of a special purpose manipulator system for the Large Space Telescope Project for the Goddard Space Flight Center.

It was to be a three-month assignment for the former native of Warsaw, who had emigrated to Canada in 1956 and, after a series of jobs, joined DSMA to do what he loved best: work on conceptual designs for projects. Tony had worked on wind tunnels, large optical telescopes, nuclear facilities and now it was to be on a key requirement for the shuttle's manipulator system.

Reporting to Nick Weyman, the Spar manager for the project, Tony joined Spar's Francis H.(" Frank") Mee, head of the mechanical development laboratory, and Mark White, a summer student from the University of Toronto, as a threesome to work on a full-time basis with Wayman and Ernie Groskopfs of Spar, spending a good deal of time with them in brain storming sessions for exploring ideas and approaches.

At the time a hard metal claw device was being considered but a revised set of concerns by NASA led to the conclusion that the claw concept might not work or that any kind of stiff grabbing device might not be effective. " In the first place, even though objects are weightless, the shuttle moves at a different speed relative to the object to be caught and the subsequent jolt on first contact could cause damage to either the arm or the object to be caught. We were, therefore, looking for a way to capture things in space that would overcome such a jolt," Tony recalls. " You could manipulate the shuttle to synchronize its velocity to that of the orbiting object but it would take too much time and fuel – things you have to be conscious of when it comes to space activity."

In a roughly hand-written preliminary report to his DSMA boss, Doug Knowles, Tony reported that, "it was concluded that all proposed systems would not fulfill the specs requirements by an acceptable degree and attention was directed to new ideas. There were many," he continued. One, however, had the imagination of at least three members and work began on Frank Mee's concept of an iris type device. He, according to Lydia Dotto's book, written to mark the 25th anniversary of Spar in 1992, came up with the concept when he took a camera apart to study how the lens worked. Tony with others then worked on how that concept could be applied to such a device for capturing objects in space.

This and other ideas by various people failed to fully satisfy the NASA specifications, however, "making the solution to the problem appearing more and more remote," Tony remembers. He admits he became frustrated and felt uncomfortable at returning to DSMA without an answer " because I felt in my bones we were almost there." His bones were right: during the last week of his assignment, he went for a walk in the Spar parking lot, with a cigarette in one hand and a rubber band wrapped around his thumb, middle and little finger to form a triangle in the other. " As I opened my fingers the triangle enlarged and when I closed them it got smaller. Suddenly I wondered if replacing the elastic with three flexible steel cables that could fit around the probe protruding from the spacecraft would provide the soft contact we were looking for."

Convincing the experts, however, was not quite as easy. " I remember standing in front of a roomful of people, space experts from NASA, scientists, engineers and others, nervous beyond description explaining the concept, showing viewgraphs and the primitive model Frank had put together, and answering all sorts of questions. They wanted to know how such flimsy cables would be strong enough, and how it would capture an object when approached from various angles. All sorts of questions, but finally they were satisfied and the three-wire crossover concept was approved." Tony's hand-written report was much more succinct. He wrote, "After weeks of making layouts, and many meetings and discussions, my snare-wire, cross-type end effector finally emerged. Preliminary positive feedback from NASA gave impetus to the idea and a full-size metal device was built." His report also concludes that he didn't remain at Spar during the next stages of its development.

"I made some casual visits to Spar to keep myself abreast of the latest modifications to the end effector model being built and tested in the Spar laboratories," Tony recalls, while work in the development, construction, test and commissioning was carried out by Frank Mee and others at Spar, and eventually led to the filing of a patent application in the United States by Francis H.A. Mee. Its preamble began, "What I claim as my invention is: 1: A grappling device comprising (a) a grappling head having a grappling window opening inwardly thereof along a grappling axis." Nearly 13 more pages of text and 15 drawings accompanied his completed document filed at Aurora, California, and the U.S. Patent and Trademark Office awarded him Patent 4,105, 241 on August 8, 1978.

Thirteen years later, on October 3, 1991, two patents were filed with the Canadian Intellectual Property Office, one described as an end effector filed by Francis H.A. Mee and Henry Liu, and the other by Mee alone for an "Interface For Orbital Replacement Units." These were granted in 1992.

Another who spent two years on loan to Spar was John White, serving as a project manager on the team working on the various joints of the arm itself. He and Peter Tillson was another at the Spar plant as they worked on the control systems for the wrist, elbow, and shoulder joints that led to the successful launch of the Canadarm in 1981.

While DSMA employees played a major role in helping build the components for the arm itself, they were also entirely responsible for developing the entire test rig required to make sure the arm, as a system, would indeed function properly in space. To do this, a team of DSMA engineers and designers spent several years working in their own laboratory at 3235 Wharton Way, Toronto, to design, build and develop the various test rigs required to assess how each of the joints must react in a weightless environment.

Four major components were developed: a systems rig to support the weight of the 907-pound, 50-foot arm to operate at one "g" for gravity ground level, the closest it could get to the near zero "g" conditions it would experience in space. This meant designing and constructing a plinth, to which the arm is attached, and three separate units to support the upper and lower arm and wrist that are mounted on air bearings. The entire assembly then would move on air pads on a specially designed floor that was built at Spar's Ormont Drive plant to be used in qualification testing and handling familiarity sessions by NASA's payload specialists that would be operating the Canadarm.

To do this, an axial test rig was designed and built to assess the effects of gravity on the various joints of the arm, and a large flywheel, designed by Ben Tripp at the ATCON lab, proved to be a practical way of providing full load simulation and making it possible to undertake the development of several of the arm's mechanical and control subsystems.

Even before Canadarm was tested in space, the Association of Consulting

The two astronauts involved in the first test of the Canadarm in space, Colonel Joe Engle (far right) and Captain Dick Truly, are shown talking to Lloyd Secord (far left) during a visit to the DSMA ATCON offices in December 1981, a month after their flight. They made the trip to thank members of the staff for their participation in the success of the project. Looking on, left to right, are Al Bullock and Doug Knowles of DSMA ATCON.

Engineers of Canada were so impressed by the test rig itself that they recommended DSMA ATCON be given an "Award of Excellence" for the project. The citation outlined the problems faced and overcome and summed up its findings for the award by stating: " From all of the above it can be seen that an innovative approach was necessary to achieve a simple and cost effective concept for a Systems Test Rig and that it was obtained by the imaginative use of a combination of three quite common mechanisms – air bearings, constant force devices and flexure pivots." Doug Knowles, as manager of the test rig project, made the trip to Ottawa to accept the award on October 29, 1980, from the Right Honourable Edward Schreyer, Governor General of Canada.

For Lloyd Secord an even more memorable event occurred on November 11, 1981, when he was among the invited guests for the launching of the shuttle *Columbia*'s second space flight, this time with the Canadarm fastened along the left wall of the cargo bay. Two days later the two astronauts in the shuttle, Colonel Joe Engle and Captain Richard Truly reported back to Mission Control, "Okay, the arm is out for the first time. She's been employed ... working great ... It's a remarkable machine and it's doing exactly what we hoped and expected." A month later they met each of the ATCON staff when NASA sent the two-member crew to visit Spar Aerospace and ATCON in Toronto, and the CAE offices in Montreal to thank personally the firms for making their flight so successful.

"Everyone in the office turned out in an excited gathering as Mr. Secord welcomed the astronauts and introduced them to the staff," reported the employee newsletter *DSMAction* as Tony Zubrzycki, Rudy Williams, Fred Hummel, Al Bullock, Tony Stonehill, Joan Smith, Joe Lehman, Dan Simec, Ziggy Akots, Della Trude, Bob Ross, Cec Corke, June Bell, Doug Ferguson, Bob Hadley, Richard Hannaby, Colin Martin, Ursula Mueller, Ed Perchyshyn and Peter Tillson were on hand to meet the first real users of their efforts. Gifts were exchanged – paintings rendered by ACTON's staff illustrator, Rudy Williams, of *Columbia* in orbit with the arm extended, and the astronauts presenting a composite plaque of the stages of the flight. "It was certainly a great moment for me and I think a proud one for all the ATCON staff," Lloyd recalls.

7 The Telescope Years

DSMA designed a unique support system for the huge mirrors in a number of telescope projects in the 1970s. Under the leadership of Lloyd Secord and John Farrell, a DSMA team created a pneumatic bellows system to prevent the sag or distortion of 150-inch mirrors weighing as much as 16 tons as they were moved from a vertical to horizontal position. Telescope projects to adopt the new support system included the Anglo-Australian telescope in the Snowy Mountains of Australia, the Canadian/France/Hawaii telescope at Mauna Kea, Hawaii, the telescope at Kitt Peak, Arizona, and its twin unit in the Southern hemisphere near Santiago, Chile. Photo: Courtesy Jeff Stringer.

As early as the mid 1960s, DSMA became involved in another major field of endeavour that was to last for two decades – the design of astronomical telescopes. As Lloyd Secord, who became the driving force in this field for DSMA recalls, "There was a flurry of world-wide activity at the time with astronomers getting newer and larger telescopes. The Americans were building a couple, the British and Australians were talking about one and the Germans, French and Italians were also interested." A European consortium (ESO) was also considering a specialized telescope to be erected at La Silla, Chile. "Even Saudi Arabia, Iraq and Iran, each with excessive revenues from oil exports, were exploring possibilities of restoring Arab leadership in stellar exploration with very large telescopes, but none materialized," Lloyd remembers.

Canada too, decided to build one on Mount Kobau, B.C., to be called the Queen Elizabeth II. It was announced during the Queen's Fall visit in 1964, and when Dr. Van Steenburg, deputy minister of Mines and Technical Surveys asked Ken Tupper, a vice-president of NRC at the time, if there was anyone in Canada might be able to design it, Tupper recommended DSMA. He described it as a sound engineering organization, "... that doesn't work from old applications and what has been done before, but starts from the ground up."

Lloyd was invited to Ottawa to discuss the Queen Elizabeth II project. He also visited Dr. Robert Petrie of the Dominion Astrophysical Observatory (DAO) in Victoria "who was quite naturally skeptical of our ability to succeed on such a major project but did give us a chance to help their in-house mechanical technician to work on the cover of their 72-inch aluminizing vacuum chamber." This work was carried out by Walter Grundmann, a young engineer with Dilworth, Secord, Meagher and Associates, and as a result, Lloyd adds, "they felt we were useful and more comfortable with our capability."

Lloyd then put together a team to carry out the work for the QE II telescope. Ernie Eggmann was to be in charge of the design team in Victoria that included Walter Grundmann and John Farrell, a native of Dolbeau, Quebec, and a 1964 gold medal recipient in mechanical engineering at Queens University. John had also obtained his Master of Science degree that September and when he applied for a job with DSMA, Lloyd asked a former classmate of his, Ken Rush, now a professor of mechanical engineering at Queens, for his opinion. "I was told to get him if you can," Lloyd recalls, "and we were fortunate to do so for he became a key engineer for us." Two other relatively new employees, Daniel Simec, of Kingsville, Ontario, and a George Brown College graduate, and Roy Fedoruk, a 1964 Ryerson graduate, were also assigned to the project in Toronto, "and we all worked diligently at concentrating our telescope learning period with DAO as our tutors." Lloyd adds.

In October 1965, a 150-inch mirror of fused silica, weighing 16 tons was ordered with an expected delivery date in 1967. The largest one ordered in more than 30 years, the contract was made that early as it would take another four to five years to grind and polish the surface. DSMA had no involvement with the ordering of the mirror or its grinding and polishing but were contracted to provide a detailed design concept for the telescope and its dome. This called for designing the telescope including the precision support system for the mirror that was scheduled for full operation sometime in 1975. The estimated cost, to be funded by the Federal government, was five million dollars.

Astronomers from the Dominion Astrophysical Observatory of Victoria, B.C., headed by an Australian, Dr. Graham Odgers, and others from the University of Toronto who operated the Dunlap Observatory that had been, with its 74-inch mirror, the world's second largest telescope when completed in 1935, were to work together on the project. Conceptual design work and numerous meetings with the astronomers and DSMA people took place at both centres. There were also meetings in California with Bruce Rule, chief designer of the Palomar 200-inch telescope and with Bill Baustrain who was heading up the design of the proposed four metre telescope at Kitt Peak, Arizona. John Farrell recalls that not all the meetings in Victoria were held at the observatory but in a local pub where Dr. Odgers, Ernie and other members of the team would have a few beers with a meal. "It had paper tablecloths and we'd cover one or more of them with drawings and concepts, sometimes staying there until closing time, always taking the tablecloths home with us. A lot of wonderful ideas came out of those sessions," John laughs.

In a paper prepared by Dr. Odgers and Lloyd Secord and given at the Congress of Canadian Engineers on June 2, 1967, Lloyd outlined some of the problem areas and concepts proposed. It included an outline of Ernie's work on the mounting system for the mirror of the telescope and its ability to accurately point to and follow faint stars within the dome-like structure, and John Farrell's conceptual design of an advanced support system required for the mirror itself. It was the latter that required the most attention since it was known that the mirror, some 24 inches thick and 150 inches across, would sag under its 16 ton weight as it moved from a horizontal to a vertical position. This sag, while tracking a particular star or galaxy, would cause a deflection "more than a hundredfold too extreme to cause a blurring of the image," said Lloyd. He then outlined how Farrell and his team in Toronto came up with a pneumatic bellows system with 24 supports on the back surface of the mirror. Essentially the bellows, each with a separate controller – a floating piston – made it possible to supply air under the sections of the mirror at the correct pressure as it moved from 0 to near 90 degrees. Lloyd also pointed out that "A prototype of the bellows-controller was built and experimentally tested to prove the somewhat radical departure from conventional systems and repeatedly performed within an accuracy of 0.1 percent."

While the DSMA team addressed the technical aspects of designing the mechanical, structural and control elements of the telescope, the astronomers at the Dunlap Observatory and the DAO were having their own series of problems, some of them of a technical nature but mainly about where the telescope should be located and what features should be included. Suggestions included a grinding facility for making mirrors and lenses, facilities to accommodate tourists in a park area with an astronomical display centre and "sumptuous quarters for visiting astronomers expected from around the world," Lloyd recalls. Mainly, however, Farrell, who sat in on many meetings, claims: "There was a real dog fight about where the telescope should be located. The astronomers from Dunlap were critical of the site in British Columbia to provide an adequate number of nights with observing conditions for such a world class facility. They wanted to have one in Chile while members of the Dominion Observatory were equally adamant it should be built at Mount Kobau." Finally it was recommended that both be built.

As a result, the Federal Government, shortly after the election of Pierre Elliot Trudeau in 1968, stepped in and said in effect, "Fine, nobody gets it," John remembers. A number of added reasons were given, one being the perceived increased costs (estimated as high as $20 million for the enlarged project concept) and the fact that such expenditures might accrue for a project to be built in Chile rather than Canada.

"However," Lloyd recalls, "all was not lost as far as DSMA was concerned. Unlike commercial enterprises which have an overblown concern for secrecy, the preliminary concept report for the telescope project got widely distributed to astronomers around the world for their comments and critique so they were aware of the work we did. That opened the door for us." It was a door that Lloyd also pushed on hard to keep open by visiting Europe and the Middle East, anywhere where there was talk of building telescopes, Germany, France and Italy in particular. "There was also talk of Great Britain and Australia teaming up to build a telescope in Australia that would serve as the foremost facility to be located in the Southern hemisphere. When a disagreement arose about whether the British or Americans should design it, we were found acceptable to both and got the contract," Lloyd recalls.

That was in 1969, and other promising contracts followed in 1970 and 1972, making DSMA "involved in the design of every major telescope project in progress worldwide," John declared. The company's newsletter backed up the claim. In July 1970, it reported, "We have now received a substantial order for the preliminary work on the French telescope." Its next edition, two months later, added that meetings in Paris covered a wide range of topics on the French 3.6m telescope and in March 1971, again reported that Ernie and John with Lloyd, "had made a whirlwind tour of European capitals to have discussions with French and Italian project officials and that in Geneva, DSMA had the privilege of presenting three papers relating to large telescope design before 130 delegates – telescope experts from around the world."

The same issue also claimed the company had a contract with the Carl Zeiss Optical Company of Oberkoken, Germany, for the design, fabrication and commissioning of the axial support system for the first of two telescopes to be supplied to the Max Planck Institute. Another item in the same issue announced that DSMA had been awarded a contract for the structural analysis of the European Southern Observatory 3.6 metre telescope, and happily added, "This brings DSMA's batting average to 100 percent for the complete design or partial involvement on all presently on-going large telescopes."

While these projects were being contracted for preliminary work, the Anglo-Australian telescope to be built at Cunabarbrand in the Snowy Mountain region of Australia, some 100 miles south east of Canberra, was underway and became the first to utilize the mirror support system developed by John Farrell and his colleagues. Prince Charles officiated at the opening ceremony of the Anglo Australian Observatory in 1974, and at the time Australia could boast that it had one of the foremost telescopes in the world. "It soon became the standard by which other telescopes were ranked," Lloyd reports. A number of other telescopes adopting the unique bellows support systems were also built over the years. They included the Canada-France telescope built in Hawaii, two supplied to the Max Planck Institute of Germany, and the University of Arizona's Kitt Peak installation in that state and its twin unit in the southern hemisphere at Cerro Tollolo near Santiago, Chile.

But a number of other telescope projects that DSMA were involved in never had an official opening of any kind because they didn't get past the preliminary stage. The Italian proposal was a good example. When Lloyd made his first trip to Europe, following the cancellation of the Queen Elizabeth II project, he met with Professor G. Rhigini of the Science Museum in Florence who, on hearing Lloyd out, admitted that they had been unsure how they could proceed with the engineering, "... but now that DSMA is available, I think we have the answer." Later, ATCON received a wire asking it to submit a proposal and Tony Zubrzycki, another of the design engineers who had worked on the previous projects with

Dan Simec and Roy Fedoruk. remembers how they worked virtually around the clock for three days and nights to come up with a concept and sent it to Florence *poste haste*. The Italians were stunned at such a quick response and agreed DSMA would get the contract, but before it went further the Italian government cancelled it.

An even more ambitious project was underway by then, due to the persistence of Lloyd when he visited Tucson, Arizona, and heard a NASA project engineer talk about launching a smaller companion to the Hubble space telescope, to be called Starlab. The official admitted it was on hold

Lloyd Secord, left, and Ernie Eggmann look at Gallileo's original telescope during a private tour of the Scientific Museum at Florence, Italy. For almost two decades, DSMA and DSMA ATCON played a role in virtually every major telescope project around the world. Photo: Courtesy John Farrell.

because of cost overruns, and while many of the astronomers present were skeptical about the effectiveness of a telescope in space, Lloyd teamed up with Gordon Walker, a faculty member of the University of British Columbia, to meet with the NASA official. They discovered he was working with Australian astronomers on a detector that would pick up photons of light from distant stars, and said it would be up to us to get their agreement to join them in any activity," Lloyd recalled.

On returning to Canada, Lloyd checked with friends at NRC and the Department of Trade and Commerce who were reluctant. "At the time, however, I happened to be chairman of the Space Committee of the Air Industries Association of Canada and suggested that I, in that role, approach the Australians directly, to which they agreed as long as I didn't commit them to spending any money." Lloyd immediately wired the Australians and got a return wire from Don Mathewson the next day, January 17, saying "... they would be delighted to have us involved."

A meeting with NASA was arranged that included Lloyd and Claus Wagner-Bartak, head of the robotic division at SPAR Aerospace Limited of Toronto. Lloyd promoted the idea that DSMA could handle the design phase with SPAR to build the telescope, Australia handle the general installation and NASA provide the shuttle launch in space and data transmission back to earth. There was interest and Lloyd and Claus jointly approached the Canadian government. As a result, the final agreement called for Canada and Australia to provide the telescope, NASA to launch it and the data communication from the telescope received at a NASA centre would be forwarded to Canada and Australia.

For the next three years the three-way partnership happily existed, taking turns to conduct scientific meetings twice a year. Then it became clear that to pursue the project Canada would have to put up $20 to $25 million. The government said: "No." Their reasons, according to Lloyd, was because the budget covering this type of science didn't have enough contingency reserve to cover such a large commitment at a time when it was exercising fiscal restraint for non-vital projects. "Canada's withdrawal was the fatal act that terminated the project," he adds.

There was one more contract in the works, however. Canadian National Railways had agreed to build the CN Tower as part of a Metropolitan Redevelopment plan then in progress. Originally, they planned to have six telescopes and three were initially installed at the 1372-foot level of the closed observation deck so visitors could buy a magnetically encoded card for a $1 to see as far as 75 miles away. DSMA won the contract to design the mount and control structures for the units. Ernst Leitz Canada Ltd. manufactured the optics and zooming features and eyepiece head, while OPCON Associates, because of the space limitations, developed the optical path directed by mirrors making the units more like a periscope. In fact they were called peritelescopes.

Keith Lacey was the project manager with Peter Tillson assigned the job of coming up with the controls for the units. "One of my major problems was devising an automatic shutter within the optical path because if someone accidentally pointed to the sun, it could burn your retina in a flash and blind you. I had such a scare when I was setting the shutters up. It was worse than a welding machine flash and I couldn't see for a good 60 seconds," Peter remembers. He also had to devise control limitations so that users could not invade the privacy of people living or working in nearby buildings because the magnification would even permit reading correspondence. On a happier note he, as a camera buff, claims: "I took some wonderful pictures of the moon while working on the project." The tower was officially opened on June 26, 1976.

However, a major problem continued. People stuffed paper and other debris in the coin mechanism, and Peter was called back to try and find a solution. "Over the better part of a year, I went up and down those darn elevators," he recalls. Several solutions were tried but it was "almost impossible to make them 'public proof,' and eventually they were removed." Some years later, sets of coin-operated binoculars were installed.

In 1978, Lloyd, with Ian Billington, went to Iran to meet astronomers there in search of another telescope project which was said to be on the verge of being awarded. Discussions broke down, however, when the Iranians insisted on changing the proposed earlier conditions of payment covering the initial design phase. "We felt what they proposed was too risky and as they were adamant, negotiations broke down," Lloyd admits. He adds with a smile, however, "perhaps it was just as well for some months later the Shah was overthrown and we would probably never have been paid for the work we would have done in the interim."

Still, the breakdown of negotiations with Iranian astronomers was a disappointment, for ATCON was finding fewer and fewer projects due to a slowdown in the economy worldwide, causing the cancellation or deferment of numerous projects at both ATCON and DSMA International where a wind tunnel in China was cancelled. At ATCON, the heavy water plant at LaPrade, Quebec, suffered a similar fate. Another personal blow to ATCON occurred with the death of Dr. Ian Billington who had joined the company in 1957 and had been vice-president of DSMA ATCON since its formation in 1978. He was widely known and highly regarded for work in the aeronautic, nuclear, wind tunnel and oil industry fields with many technical papers to his credit. He died of leukemia at the age of 57 in 1981.

In 1984, another small but interesting contract for a telescope came about when Saudi Arabia, which had been visited several times over the preceding 15 years with no success, had a scientific team covering many disciplines, visiting the U.S. and Canada. "By luck, the Saudi tour leader knew of our previous attempts and asked that we be included at a luncheon meeting," Lloyd recalls. "The Saudis said a major telescope for leading edge science was no longer in their plans but wondered if we could help them with a current matter of immediate concern. The start of the next Islamic year was traditionally determined by the first human sighting of the new crescent moon. They asked us if we thought it was possible to telecast over the Islamic TV networks, a split screen view of the sun setting simultaneously with the appearance of the new moon following it so that Saudi Arabia could authenticate a single sighting applicable to all the Mid-East."

While not committing himself on the spot, Lloyd promised to a reply within a month, "... as they wanted it by the following May to coincide with Ramadan, the start of their annual religious pilgrimage to Mecca and it was now August 1984." Despite the tight schedule, it presented Lloyd with the kind of engineering challenge he liked, and he agreed to undertake the project with the added caveat that if the date was met, there would be a $50,000 bonus added to the contract.

There was no time to design a telescope from scratch, nor fortunately was there a need to, as commercial telescopes for the amateur astronomer could be modified with special instrumentation fitted to them that would work. The concept was to link electronically in Siamese fashion, two telescopes that would follow the setting sun and the rising moon which was invisible to human eyes because of the low light output from a slender crescent and lingering sky brightness of the sunset. The two were purchased and the coordination function to determine how to position and coordinate the telescopes to capture setting sun and the faint crescent moon was studied by Peter Tillson, assisted by Don Waddington and Dan Simec who had worked on earlier telescope projects.

Shakedown testing of the controls was conducted on several cold spring nights on Dan Simec's property far enough north east of Toronto to escape the city glow and, in early May, the team consisting of Don Waddington, Dr. Don Welch, an astronomer from McMaster University, and Dr. Andrew Robertson set up camp in the desert some 30 miles from Hail, Saudi Arabia, where several more days were used to install the two telescopes in an enclosure built by the Saudis. "Unfortunately," Lloyd recalls, "the builders misunderstood the reason for an opening in the north wall and made the opening on the west wall so that we could not see the North Star from the observation floor. That meant we could not quickly align the axes of the two telescopes using that star and had the tedious job of finding suitable bright stars near the zenith whose position was known precisely and visible through the western wall opening." This was done, one day before the Ramadan observation," he adds, "through the perseverance of Doug Welch, who had been retained throughout the project for astronomical guidance."

With that accomplished, the telescope and observatory building were put to bed for the night with a "euphoric feeling that we had met the Ramadan deadline," Lloyd recalls. In the morning, however, it was discovered the telescopes were no longer correctly aligned. "No one knew how it happened, but we knew it would be impossible to realign them for that evening as it had taken us several days to do so previously." Lloyd adds.

While the chance had passed, the Saudi officials were satisfied that the team had fulfilled its obligations and feted the Canadian team and the student astronomers assigned to the project with a traditional desert feast on rugs spread over the sand. They also paid the $50,000 bonus before Lloyd, Doug and others returned to Canada, never to discover if the two telescopes were ever used again to try and confirm the correct day for the start of Ramadan.

One reason why the outcome was never discovered was because ATCON went out of business early the following year. The events leading up to the end of its operations started in January 1986 when a routine meeting with its bank, ostensibly to review the company's financial statements of the previous fiscal year, turned out to be the venue to raise the issue that the bank held no personal guarantees. Lloyd pointed out that over the past 30 years that issue had been raised from time to time but never pursued, but a few days later Lloyd received a further request that shareholders collectively provide $150,000 in personal guarantees.

On February 8th, Lloyd and 17 employee shareholders agreed to refuse such guarantees, which prompted the bank to issue, ten days later, a letter giving formal notice of the appointment of Price Waterhouse to act as Agent for the bank and collect from the collateral provided whatever it could of the debt owed to the bank. The amount was $423,795.61. The locks on the doors of 4195 Dundas Street West were changed as were those at its laboratory, ATCON Services Limited, where equipment being manufactured for Ontario Hydro was also impounded. DSMA ACTON ceased operations.

8 Wind Tunnels – At Home and Abroad

It is no accident that Canada is a pioneer and leader in wind tunnel development. As mentioned earlier, Rupert Turnbull built one out of old crates at his workshop in Rothesay, N.B., and before another decade had elapsed the National Aeronautical Establishment (NAE) of the National Research Council (NRC) had constructed a primitive one in an old factory on John Street, Ottawa. In the early 1940s, two more were built within the NRC's J.B. Parkin building which is also known as M2, at 1500 Montreal Road. Then in the mid-1950s it was decided to build what can be described in several terms that include an Intermittent Trisonic Blowdown Wind Tunnel at Ottawa's Uplands Airport.

Intermittent meant it could only operate at any one time for six to 20 seconds before the pressurized air ran out of its three major tanks and a refill was necessary. Trisonic because the wind speeds from subsonic, through transonic and up to supersonic (Mach 4.2) could be reached, while the term blowdown refers to the fact that the high pressure air stored in three huge 157-foot tanks when released, "blew down" through a flexible nozzle into the test section where the object being tested was in place. A principal reason for building the tunnel was to test the Avro *Arrow* supersonic jets then being developed under the direction of Jim Floyd at the A.V. Roe aircraft company at the time.

While the earlier wind tunnels were designed and built primarily by the staff of the NAE as a branch of the NRC, it was felt the new blowdown project would require additional engineering and construction expertise

A solitary worker is silhouetted against the shell of the giant settling chamber that has a diameter of 82 feet, one of the components of the low-speed wind tunnel built on a site at Uplands Airport, Ottawa, in the mid-1960s. Originally intended to test V/Stol aircraft, its two test sections of 30 feet and 14 feet have been used to assess not only aircraft but also buses, trucks, automobiles, racing cars and even a model of a submarine. The facility is still in continual use today.

and, as mentioned earlier, Dilworth, Secord and Associates and Dilworth-Ewbank were eventually awarded the contract. The end result was a wind tunnel with a number of new features jointly developed by the NAE/NRC and the two consulting firms that produced one of the most sophisticated blowdown wind tunnels in the world.

While the primary reason for building wind tunnels is to simulate wind and weather conditions on everything from space and aircraft to trains, buses and automobiles, each wind tunnel must be designed and built individually to meet the various needs of clients. An example of a particular need studied by Aiolos Engineering in 1995 was an "icing wind tunnel" near Naples, Italy, to test the effect of ice on aircraft wings and engines. It was designed primarily to test icing conditions at temperatures of -50°C whereas other tunnels have been or are being built to assess the effect of wind alone, or wind and climatic conditions. Those built to assess wind and climatic conditions are equipped with hot and cold chambers that can produce temperatures ranging between minus 40°C to plus 50°C or the effect of rain and snow on various components of the object being tested. More recently, acoustic tunnels are the order of the day, designed to assess noise factors caused by both internal and external design of aircraft, trains and automobiles.

The original blowdown wind tunnel at Uplands, conceptually designed by the staff of NAE/NRC, was enhanced at the detailed drawing stage by some innovative designs by DSMA and several other consulting firms involving in total some 100,000 engineering man hours according to a paper published in the August 1961 issue of the *Engineering Journal of Canada*. The paper, prepared by Paul Dilworth, Ken Tupper and L.A. Jenkins as project engineer for Dilworth-Ewbank, reported that the wind tunnel's 5ft by 5ft test section "... is one of the largest and most versatile

Uldis Kiesners tests the large pressure regulating valve behind him. The valve was designed by Ernst Eggmann for the blowdown wind tunnel built at Uplands Airport, Ottawa, in the late 1950s to make it possible to release wind at velocities as high as Mach 4.2 in less than a second.

of its kind" that will permit testing "throughout the subsonic, transonic and supersonic speed range from Mach 0.2 to 4.2." It also claimed that, "During a blowdown, air mass flow rates approach three tons per second and kinetic energy flux in the working section is upwards of 280,000 hp, making it one of the "relatively few in number anywhere, and in Canada, unique."

Ernie Eggmann is given credit for much of the design of the flexible main nozzle and a huge new valve that could precisely regulate the mass flow from the tanks of compressed air at fluctuating speeds. The main nozzle made it possible to move the flow of air from the tanks from Mach 0.2 to Mach 4.2 and the valve could establish the air to full Mach 4.2 force in less than 7/10 of a second making it possible to test an object in the test section for 5.3 of the six seconds it takes to void the three tanks of compressed air. The typical energy expanded during a supersonic blowdown lasting six seconds approaches one million HP, claims a project sheet on the tunnel's capabilities.

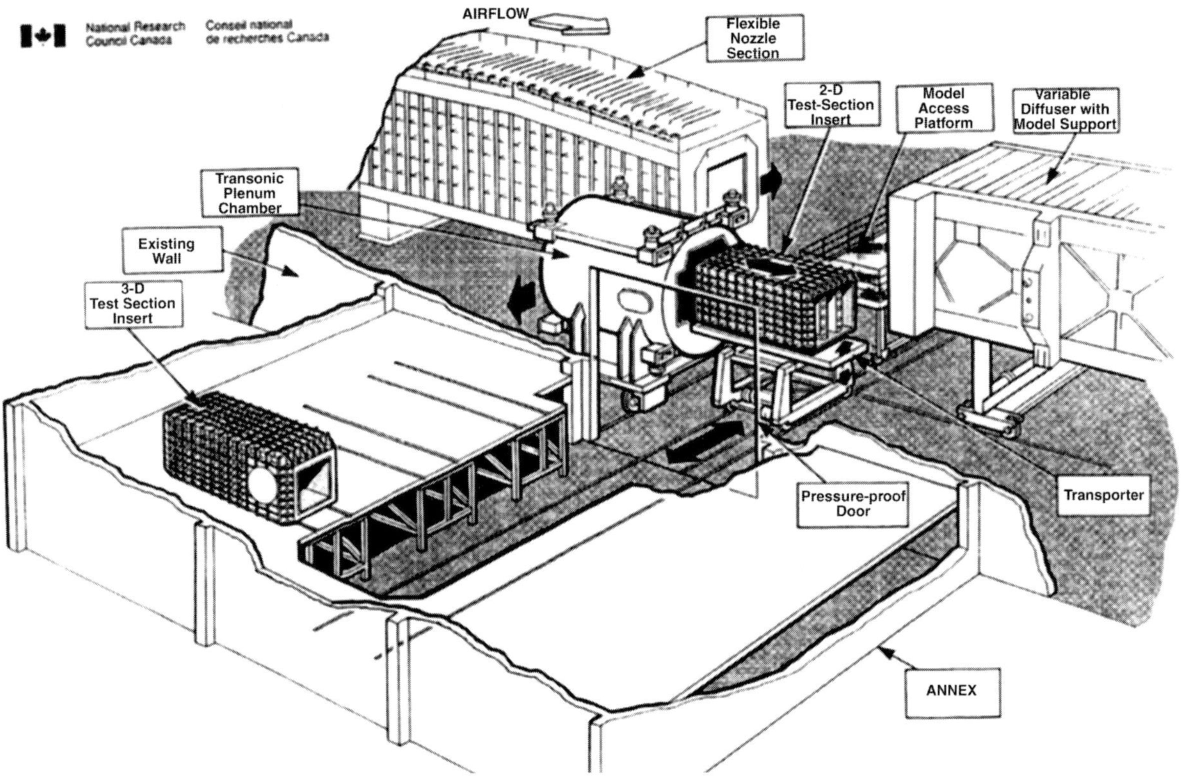

In 1985 DSMA was commissioned by the NRC to add two roll-in, roll-out transonic test section inserts to the original blowdown wind tunnel at Uplands Airport. A huge transporter on rails was built to move the new two-dimensional or three-dimensional test section insert into the existing 1.5m X 1.5m test section in a matter of two days instead of the three weeks required previously to change the configuration of the test section. The new units also incorporated the latest developments in wind tunnel technology, such as perforated walls with slanted holes to provide better wind flow at sonic speeds, and splitter plates inserted in each of the holes to greatly reduce noise levels.

Its uniqueness was also assessed in a number of other ways. During the short test periods, mass flows had to be regulated to 0.1% tolerance, the model support mechanisms, called stings, had to be positioned accurately to 1/4,000 and data measurements on the model reactions and operating parameters had to be made to an accuracy of 1/1,000.

For the next 25 years, thousands of tests were carried out in the unit, so many, in fact, that in 1985, it was decided to increase its capacity and use by adding two roll-in, roll-out transonic test section insert systems. This was done by adding an extension to the building as a storage area and a transporter on rails which could move either a two dimensional or three dimensional test section insert into the existing 5 x 5 foot (1.5m X 1.5m) test section. It had been possible to adjust the existing 5 x 5 foot test section but the down time to do so was roughly three weeks whereas the two new roll-in, roll-out units could be put in place in two days. Besides that, the new test section units featured the latest developments in wind tunnels such as perforated walls with slanted holes to provide better wind flow at sonic speeds and with splitter plates inserted in each of the holes to greatly reduce the noise in the test section.

In a report prepared on completion of the project, Lars Ohman and other staff members of the National Research Council traced the history of the development of the new units from the conceptual design, done in 1981, to the finished job in 1989. DSMA International was awarded the job of the final design in 1985 and, in 1987, became the turnkey contractor to complete the job in 1989.

DSMA International, by 1989, had built a worldwide reputation as the designer and builder of wind tunnels in more than a dozen countries, starting with India in the early 1960s. Impressed with the blowdown wind tunnel at Ottawa, the government of India signed a contract with DSMA to design a similar blowdown wind tunnel to be established at its National Aeronautical Laboratory in Bangalore. While closely resembling the Ottawa project, it called for a smaller test section of 4 x 4 feet instead of 5 by 5 foot and four rather than three storage tanks that were 93 feet longer than those used at Ottawa. Its maximum velocity would be Mach 4, and with its larger air storage tanks, its run time would be between eight and 60 seconds. The Indian government also wanted it built for them but as DSMA were not at the time ready to do turnkey projects, it arranged for Canadian Vickers to be the prime contractor.

Having won the contract, George Meagher invited Doug Hunter, an old classmate at McGill University, to join the company with the idea that he might become project manager for the wind tunnel at Bangalore. Hunter, who had graduated from McGill with George in mechanical engineering in 1942, and had spent 18 years with Dominion Bridge Company which was going through a reorganization phase at the time, decided to visit DSMA just before Christmas 1963. "I liked what I saw and decided to make the move," Hunter remembers, and was soon on his way to Port Alberni, B.C., to look into the reasons for the bearings seizing up on a paper machine. "I was working on some statistical analysis to figure out the problem when Ken Tupper, then vice-president of engineering at NRC who was also asked to find a solution, made the observation that 'Statistics are just like garbage. You had better know what you are going to do with them before you start collecting them,'" Doug recalls with a chuckle.

By the end of 1964, Doug was named an Associate of DSMA and appointed design and then project manager for the wind tunnel at Bangalore. For the next few years he made periodic trips to the site to talk to some of the DSMA and Canadian Vickers people stationed there. "One of the problems

Thousands of tests have been carried out on models of aircraft, full scale cars and trucks and numerous other projects at the National Research Council's low speed wind tunnel completed at Uplands Airport, Ottawa, in the mid-1960s. One of the more unique tests involved testing this model of a submarine for the Department of National Defence and another involved a dragster racing car in the 9m X 9m test section.

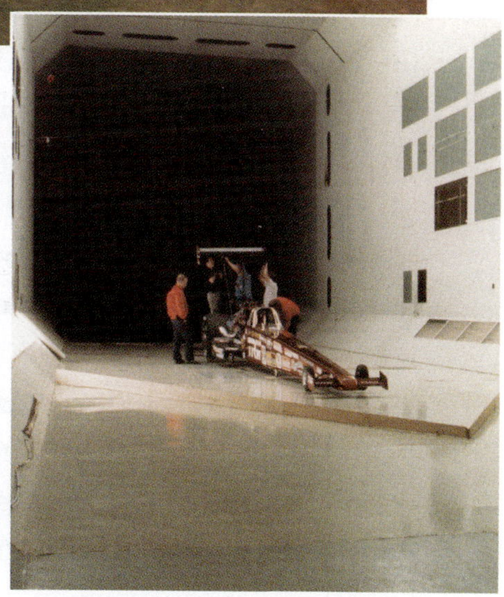

I had of course is that I didn't know much about aerodynamics or the designing of software. When I'd ask them how long it was going to take or was it within budget, I'd be told 'as long as it takes' because they didn't know either in some cases. It was typical of DSMA management, you got an assignment and it was up to you to get the job done."

Doug did get it done with George Bremner a graduate aerodynamicist from the University of Toronto who was sent to commission the highly successful job in 1967. By then DSMA had their second major challenge in Ottawa when the NAE and NRC decided to build a new low speed facility to test Vertical and Short Take Off and Landing (V/STOL) aircraft which were in demand by both the civilian and military establishments and were being produced by De Havilland in Canada and other aircraft companies around the world.

The new wind tunnel called for a whole new set of design parameters. It would be a continuous flow wind tunnel, capable of wind speeds of up to 200 feet per second. Instead of huge tanks storing compressed air, the wind would be driven by a fan measuring 38 feet in diameter. This would be driven by a 9000 hp DC motor, "with a short term overspeed capability of 14,000 hp,"according to another article in the *Engineering Journal*, written this time by George Meagher, who was project manager, and Ian Billington in the November, 1967 issue. Its test section would be huge compared with the earlier blowdown tunnel, measuring 30 by 30 feet and 75 feet long, or in metric terms, 9 X 9 metres square and 22.5 metres in length. The test section was also designed so that it could accommodate a smaller test section with a 14 by 14 foot opening to permit wind speeds up to 475 feet per second. While there were some larger wind tunnels in the United States, the new NAE structure was the largest under construction at the time.

Design of the fan was a major component and once again Ernie Eggmann played a prominent role as its designer, a role he was to fulfill for a number of low speed continuous flow wind tunnels. The fan consisted of eight blades roughly eight feet in length and mounted in a hub of 22 feet. The blades, essentially made of an aluminium sandwich construction, could be adjusted to serve both the 30 and 14 foot test sections. The fan could be controlled to rotate at speeds of 24 to 244 rpm. Rodney Bell recalls Ernie's working habits when it came to designing fans. "He would pore over old Swedish books on fan designs and come up with the most sophisticated and fabulous fans. He really was a genius when it came to creative engineering."

With two blowdown wind tunnels and the successful launch of the low speed facility at Ottawa to their credit, DSMA, by 1968, was in a position to offer their services worldwide and did so under the aggressive salesmanship of George Meagher and George Bremner, but it sometimes took years before a decision to build a project was made. Two examples of the lengthy times intervals are the Supersonic Indraft Wind Tunnel for the Defence Research Establishment, Valcartier (DREV) in Quebec and a low speed wind tunnel for Holland. Both started in 1968, the Indraft wind tunnel to be built at Valcartier with a 2 foot by 2 foot test section and the preliminary design for a low speed wind tunnel for The Netherlands. Eight years later, the DREV project was commissioned and four years after that the preliminary study for The Netherlands turned out to be the DNW wind tunnel, a joint project of both Holland and Germany that is still considered one of the finest ever built.

An indraft wind tunnel is opposite from a blowdown wind tunnel in that wind is sucked instead of blown through the test section. Very dry air, initially stored in a collapsible bag is drawn through the test section by a

The photo above shows Gerrit Zorgdrager as the project engineer (red helmet) watching with an official of the DNW wind tunnel in Holland as a section of the outer shell is put in place. Another feature of the interior of the DNW wind tunnel is the turning vanes that direct the wind flow at one of the 75 foot high corners. The black streaks between the panels are rubber seals to ensure that no air escapes. In 2001, Gerrit, as vice-president and general manager of Aiolos, was a guest at the 25th anniversary celebrations of the start of the project that is still considered one of the finest wind tunnels in the world. Photos: Courtesy John Farrell.

huge globe-like tank evacuated of air, to achieve subsonic through supersonic air speeds. Less expensive to build than blowdown wind tunnels, funding problems were, nevertheless, one of the reasons for the delay. Over that many years, a number of DSMA International people became involved in the job, including Carl Balkwill, Gerrit Zorgdrager and Kochi Nishiwaki, who are now employed by Aiolos Engineering. A second indraft wind tunnel was built by DSMA International for Fuji Heavy Industries at Utsunomya, Japan, in 1980, the same year that the DNW low speed wind tunnel was completed in Holland.

Uldis Kiesners was to manage The Netherlands' project that, after submitting the preliminary concept, went in limbo for three years. Then DSMA was asked to design a fan for a pilot model wind tunnel. This was ready in 1972, but another two years elapsed before another contract was signed for the actual wind tunnel with an 8m X 6m test section.

DSMA engineers and design staff had hardly started working on this project when they were summoned back to their drafting boards as The Netherlands and German governments decided to make it a joint undertaking and wanted more test sections and numerous other features included in the project. As a result, the low speed wind tunnel, known by the initials DNW became, and still remains one of the finest wind tunnels existing in the world today, a claim made by John Farrell, who became the site and project manager, and echoed by Gerrit Zorgdrager as project engineer, both of whom spent more than three years on site at the North East Polder, about a 30-minute drive north of Amsterdam.

Construction of the DNW project got underway on July 1, 1976, when Ministers of the Dutch and German governments drove a concrete pile about 10 metres long into the ground before some 300 guests, including

George Meagher, Uldis Kiesners and Gerrit, who cites a number of reasons why it was, and has retained the claim of being, one of the world's premier wind tunnels.

"It has one of the finest air flow qualities of any wind tunnel in operation anywhere," says Gerrit, "and has features that make it one of the most diversified ever built," John Farrell agrees. Its test section configurations, for example, are the original 8m x 6m unit redesigned to also accommodate a 6m x 6m unit and a 9.5m by 9.5m solid wall test section that can be converted to what is known as an open jet test section. Maximum wind speeds created by the largest fan made for a wind tunnel up to that time

Later, DSMA experts came up with a solution to move a 300-ton section in and out of the test centre: large, air pads making it possible for four people to push the section into place easily. The photo shows similar application of air pads to the NLR 3m x 2m wind tunnel, also designed by DSMA. Photo: Courtesy NLR.

range from 60 metres per second in the 9.5m x 9.5m unit to145 metres per second in the 6m x 6m test section.

With three test sections to accommodate, Gerrit as project engineer was confronted with the problem of moving them around inside the steel and concrete building housing them. "One of the sections was about 300 tons and measured 17 and a half metres high, 17 metres wide and 25 metres long and at first we thought about using rails. But rails meant the logistics would have been prohibitive – in fact a lot of the moves would have been impossible to make – so we came up with doing it with air bearing pads. It hadn't been done before, but we thought it could be done. I remember," Gerrit adds with a hearty laugh, "when Roy Fedoruk and I were sitting in the office looking out of the window at an office building across the street and Roy saying, 'You see that thing across the street? That's what we have to move.'" The air pads, however, worked, making it possible for four people to push the 300-ton section into place when necessary.

The huge fan, measuring 12.30 metres in diameter and the biggest built to date for a wind tunnel was another challenge for engineers of the two countries, and Ernie Eggmann once again served on the fan design team. John recalls that the fan blades were made of 10% fibreglass and 90% carbon fibre. "We started with the ratio the other way around because carbon fibre is extremely expensive but when we finished up studying the deflection of the blades and other factors it meant using the 90 to 10 ratio in favour of carbon fibre." A 10,000hp motor made it possible to rotate the blades at 244 rpm. "At that speed the blades actually stretched so that they virtually touched the outer rim of the fan but that gave us a minimum of air leakage and maximum fan efficiency" John claims. At top speed the concern was the vibration factor and the noise level generated, the latter overcome by introducing acoustic lining at key areas of the wind tunnel to make it the first that could be described as an aero-acoustic facility.

It was also one of the first to have a moving belt ground plane, measuring 7m long and 6m wide to replace the standard floor section in the 9.5m x 9.5m unit which could reach speeds of 216kph making it "... a valuable tool to execute correct tests at ground level," a booklet produced about the completed wind tunnel reports. Another feature of the project completed in 1979, included a scoop and exhaust pipe system so that jet and rocket engines could be tested in one of the sections. Designed by Uldis Kiesners and Paul Straznicky, now a professor at Carleton University in Ottawa, it could follow the engine in either a pitch or yaw position to remove jet or rocket engine exhaust gases from the tunnel with a backup plan of intake and outlet hatches to exchange the air if it should become polluted because of the engine tests."All these features still makes it one of the world's most diversified facilities," Gerrit claims and calls it "... one of the most satisfying projects I have worked on in my professional career." The 12-page booklet, produced in 1979, succinctly lists its capabilities on the second page as follows:

- the development of aircraft within their low-speed range
- the development of high-lift devices including V/STOL aircraft
- the investigation of airframe and engine noise
- aerodynamic investigations on rotors and complete helicopter models
- investigations on engines
- non-aeronautical investigations such as tests on full-scale automobiles/ trucks, building complexes etc.

The fact that it could be utilized to aerodynamically test automobiles was a feature the sponsors added in case there were times when there was a lull in aircraft testing. As a result, Volvo and Saab of Sweden and Mercedes Benz of Germany tested their models in the DNW wind tunnel until they built their own facilities. Automotive projects at DSMA began in the late 1960s when Chrysler Motors of Michigan built a multi-purpose facility that became operational in 1971. It was followed by other aerodynamic and climatic facilities for Ford in both England and Germany, Volvo in Sweden, and the British Leyland Group in England. Oil companies too were using DSMA wind tunnel expertise to test oil and gasoline consumption under various aerodynamic and climactic conditions, the Amoco Oil Company, near Chicago, and Esso U.K. being among the first to engage DSMA for such projects.

9 Wind Tunnels for the Automotive Industry

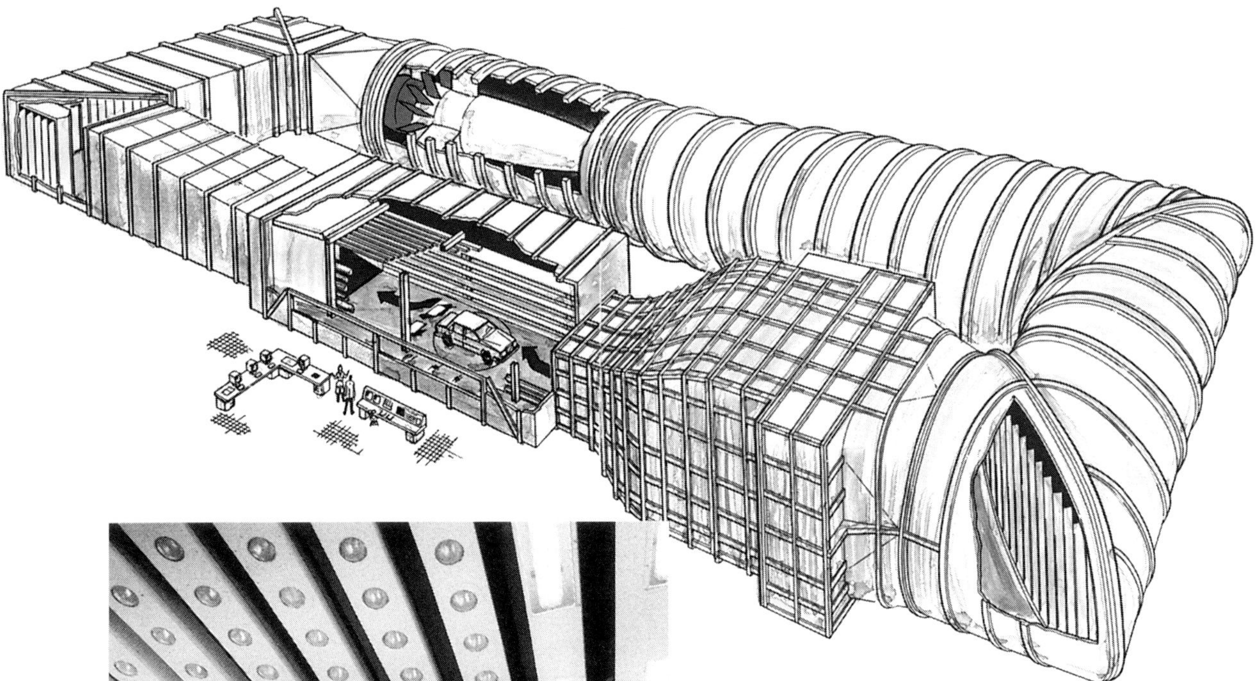

By the late 1960s, DSMA was anxious to expand their expertise to the automotive industry but they were not too successful at first. "I think we were seen as too long-haired for the more mundane wind tunnel needs of that industry," Paul Dilworth suggests. Eventually, however, Chrysler Corporation employed DSMA to engineer a multi-purpose wind tunnel at their Chelsea plant in Michigan, and by the time that was completed in 1971, a preliminary study was being carried out for Ford of Germany and the first turnkey wind tunnel project was underway for the Volvo Car Division at Goteborg, Sweden.

Paul gives much of the credit for the Swedish contract to Peter W. Gooch, who had been hired by George Meagher in 1964 to become president of DSMA's construction subsidiary, Champlain Power Products. A former vice-president of Canadian Vickers Ltd., Gooch, who had a remarkable memory and skill at learning languages, negotiated the contract with the Volvo company that called on DSMA to do the design and engineering studies and Champlain Power Products to oversee the construction of the

Left: Volvo of Goteborg, Sweden, has one of Europe's state-of-the-art aerodynamic wind tunnels designed by DSMA International and built by Champlain Power Products. It was the first DSMA wind tunnel built with slotted walls to test automobiles. It also had a fan (right) with 11 blades, eight meters in length capable of generating wind speeds up to 200 km/h, and lighting (left) in the ceiling of the test section to test the vehicle's air conditioning at temperatures to +60 C.

project that was to become one of the most sophisticated climatic automotive wind tunnels built to date.

Because Volvo was the first turnkey project for an automobile wind tunnel, the company's own newsletter to employees, *DSMAction,* carried numerous items about its construction. In March 1970, it proudly reported that the contract was won against strong competition, including the German company that had built the Volkswagen facility. Three months later the newsletter showed a photo of a model wind tunnel, designed on a scale of one to ten by DSMA, with smoke being used to trace the wind flow on the clay model of a Volvo P1400 passenger car. The model wind tunnel, designed by Pat Clark and built at the Champlain laboratory under the direction of manager Cecil Corke, was retained at the laboratory as a pilot model for many future wind tunnel projects. The same issue also reported that DSMA had been asked to design the moving chassis dynamometer system to accommodate a large range of vehicle sizes under the direction of Uldis Kiesners. While DSMA didn't make the dynamometers, over the years it designed a number of improvements which were passed on to the manufacturers to meet the wishes of its clients.

In September 1970, the newsletter reported that the Volvo building to house the wind tunnel was under construction. Other references to the Swedish project reported that the fan rotor blades, based on a concept developed by Ernie Eggmann, were completed at a factory in Cambridge, Massachusetts, that George Bremner was appointed project manager, and that Al Bullock moved to Sweden as site manager. Carl Balkwill, a University of Waterloo graduate in Electrical Engineering who joined the company in 1969, also worked in Sweden on the electrical controls. The March 1972 issue indicated that the chassis dynamometer had been installed and that Pat Clark had replaced George Bremner to complete the commissioning of the facility.

Volvo also proudly produced a booklet following its completion. It describes how its cars, trucks and buses are exported "to almost all the countries in the world. Therefore," it continued, "the climatic testing plant is dimensioned to meet all climatic conditions regarding temperature, humidity, and solar radiation for the most important markets. Air velocities of up to 190km/h can be reached on full scale cars," it added, and that, "the complete plant consists of two cold rooms, one climatic wind tunnel and also offices, conference and customer accommodation."

Wind was created by a fan with laminated wooden blades measuring 3.8m in diameter that could reach up to 620 rpm to generate the wind velocity. Two nozzles were made for the test section: one 3.2m X 3.5m used for testing trucks and buses at speeds up to 95 km/h and a smaller one for automobiles that could reach a maximum velocity of 190 km/h. The test section itself measured 14m x 6m x 4m, and "... was more heavily built than the rest of the building to accommodate trucks weighing up to 35,000 lbs."

The test section was also equipped with a chassis dynamometer made by Siemens of Germany that could create speeds up to 200 km/h, and the roof of the chamber had 246 infrared lamps of 375W each to quickly create a heat level of 68 degrees when required. In the control room, some 140 measuring points recording temperature, pressures rotational speeds, velocities, torques, humidity etc, were connected for on-line acquisition to the computer which was state-of-the-art for facilities at that time.

By the time the wind tunnel was commissioned by Pat Clark, in 1972,

DSMA was deeply involved in another climatic wind tunnel for Ford-Werke in Germany and a cooling cell for Ford in England. Then Amoco commissioned DSMA to design and construct a small environmental wind tunnel to assess fuel characteristics during the acceleration and deceleration of automobiles. George Bremner as project engineer for the Amoco facility built at Napierville, Ill., near Chicago, decided there were cost savings if the tunnel was built vertically rather than horizontally at ground level. John Farrell became project manager and remembers one contractor. "I won't say he was a fly-by-night operator but he didn't follow the rules and tried to avoid getting permits for oversized loads so he would bring the large tunnel sections at night along back roads. One day the police showed up with a citation as they were able to follow the scrape marks caused along the back road from the plant where it was made, to our site."

He also recalls that the job was quite badly underestimated and a "sort of bottom of the barrel" contractor was unloading the brine – trichlorethylene. In pumping this fluid out he didn't use the right kind of hose, and it softened and burst, pouring half a load of the brine into the basement. "You can become unconscious very quickly if you breathe it in and the foreman scrambled down into the basement to see if any of his men were there. When he didn't show up I grabbed an oxygen tank and went down and found him out cold on the floor. I picked him up, but by the time I got to the top of the stairs my oxygen had run out and I fell through the door into the parking lot where the foreman, now revived, found me lying there unconscious."

Shortly after the completion of the project, John and his wife Stephanie were in Germany. He was there as resident engineer for the Ford-Werke project at Merkenich, on the outskirts of Cologne. Bigger than the Chrysler and Volvo wind tunnels, it was to become the largest automotive test facility designed by DSMA to date. By June 1974, John reported the project was well underway but not without its challenges, arising from "material shortages, delivery delays, the oil crisis and the co-ordination problems associated with a complex hierarchy of contractors and sub-contractors in excess of 20." By 1975, Pat Clark had commissioned it and a year later, DSMA received its third award of merit, sponsored by the *Canadian Consulting Engineer* magazine and the Association of Consulting Engineers of Canada. The previous awards had been presented for the NAE/NRC low speed wind tunnel at Ottawa (1971) and the TRIUMF Cyclotron (1973).

By the mid-1970s DSMA had a number of contracts on the drawing boards, particularly in the U.K. There were two projects for British Leyland, first a hot room facility, then a cold driveability chamber and finally a climatic wind tunnel. There was another climatic wind tunnel study for the Motor Industry Research Association in Great Britain and an automobile and truck climatic chamber for Esso U.K. On the continent, the Blackstone Corporation in Sweden had authorized an engineering study for a radiator wind tunnel and Ford-Werke of Germany had signed a contract for the engineering of a pilot aerodynamic wind tunnel to test models at 1/6 scale.

By the time some of these projects were completed in 1978, DSMA had been restructured as two companies, the DSMA International organization headed by George Meagher as president and George Bremner in charge of marketing. Both men, entrepreneurial and sales oriented, placed almost all of their efforts in garnering wind tunnel and test facility business in the automotive field. In the first five years, DSMA International added at least 20 projects that included preliminary studies, engineering studies or services and turnkey projects in Canada, the United States, Great Britain, Germany and Sweden.

The largest project of that five-year period was a turnkey job for the British Leyland Company (BL) in England that was to include a large environmental automotive facility similar to the one in Germany and a climate chamber. Once again a turnkey project with Champlain Power Products (CPP) as contractor, the job brought together several newly employed people. Two of them were Robert Stapells to act as program manager for CPP, who later became its president, and Roland Vander Straeten, a native of Belgium who became president of Aiolos Engineering Company after he and other former employees of DSMA established the company in January 1994.

Roland, a graduate in mechanical engineering at the University of Ghent, came to Canada on his honeymoon in 1971 to take a degree in aeronautical engineering at the University of Toronto, with the thought of returning home. "But three things helped make the decision to stay in Canada," he likes to recall. "We liked it here, and after paying tuition fees and living in Canada for a year we didn't have the money to go back to Belgium." He went to Vancouver as head of aerodynamics for a group developing a six seater amphibian aircraft but the project was scrapped a year later and he returned to Toronto to work for DeHavilland. In 1977, he decided to take a business degree at York University, Toronto, and joined DSMA International in 1978, after receiving his MBA. Only months later, he was sent to England as design manager for the British Leyland wind tunnel, and, when things fell behind on the project, he was named site manager.

Uldis Kiesners, who began doing some research on the history of DSMA just before his death in 1998, briefly remembered that Roland was named project manager of the British Leyland job and that Fred Bauer who had been an early employee of DSMA before becoming their representative, in 1972, with offices in Switzerland, was the designer and manager of the smaller climate chamber for the same project. When Fred left before its completion, Roland became site manager for it as well. "I learned a lot on those projects," Roland said, and it stood him in good stead when he was named site manager for another major turnkey project for Porsche in Germany. "When I was named the site manager for that job some months before I had to move to Germany, the Porsche project manager questioned my ability to speak German, pointing out that the meetings would be held in that language. I spoke Dutch which helped, but not German at the time. I promised him I would speak German in the six months I had before I went there and I did. I took lessons and studied hard every week."

The aerodynamic wind tunnel for Volvo opened in 1986 with a unique feature developed by DSMA International – a slotted wall test section. Porsche was the first to agree to use such a wind tunnel followed closely by Volvo, but when the Volvo project was commissioned months before the Porsche facility, it became the first successful user of a slotted wall wind tunnel for the automobile industry. Until then, automobile wind tunnels were designed with solid wall test sections, where the air had to squeeze past the vehicle being tested or open jet facilities which meant building extremely wide test sections that allowed the air to flow more realistically around the vehicle.

The man behind the slotted wall concept was Bill Rainbird, a New Zealander by birth, who before coming to Canada in 1958, was teaching aerodynamics in England and was serving on a number of committees and commissions in Europe as an expert in that field. Hired by the Defence Research Board when it was decided to build the high speed blowdown wind tunnel in Ottawa, Bill later worked at the National Research Council with Herb Kelland and others on its construction and was responsible for its calibration and commissioning in the early 1960s. He remained with the NRC, building up its aerodynamic capabilities as well as devising a training program for Indian nationals when DSMA, with Canadian Vickers successfully won a contract to build a blowdown wind tunnel at Bangalore, India, in 1964. In 1966, Bill became Chairman of the Department of Mechanical and Aerospace Engineering at Carleton University, Ottawa, and served as a consultant to DSMA before joining them in 1977.

It was a coup for DSMA as Bill, besides having years of experience as an aerodynamicist, was a highly regarded member on numerous international committees in Europe, the United States, and the British Commonwealth. As a professor, he was also interested in establishing a research and development department at DSMA, and, with NRC support, did so with a number of bright young graduate students hired to work in it. Among them were two present-day Aiolos staff members, Peter Waudby-Smith and Sergio Raimondo whose early assignment was to work on a pilot model of a slotted wall wind tunnel for aerodynamic automotive testing.

"It wasn't a new idea," Bill quickly admits. "Slotted wall tunnels were originally developed for testing aircraft. An experimental one for automobiles was tried at CNAM, St. Cyr, France, but it was not entirely successful." Bill, however believed the concept was feasible and after several years of testing in its own small-scale pilot laboratory and further research at the NRC facilities in Ottawa, DSMA International suggested its use to Porsche and Volvo. Both companies, however, were cautious. They wanted it for their new wind tunnels but took the conservative approach and insisted the test sections should be a few metres larger than what was recommended, ending up with Porsche having a test section of 22 square metres instead of the proposed one of 18 square metres and Volvo opting for one 27 square metres instead of the recommended 23 square metre facility. Once built, however, both companies found that the slotted wall design had not only reduced the cost of building the structures but had enhanced the test results themselves by providing better instrumentation, better data handling, better flow quality and better methods for calibration and commissioning. In fact, most automotive aerodynamic wind tunnels designed by DSMA International from then on had test chambers with slotted walls.

By the late 1980s, DSMA International had completed, or was deeply involved in, a number of new projects for the automotive industry as well as upgrades of some of the previous projects. There was a Controlled Automotive Test Facility for Exxon Research and Development Corporation in Texas, a Low Temperature Durability Test Facility for General Motors and a Climatic Wind Tunnel and Heavy Truck test cell for Ford, both in Michigan. Other companies asking for upgrades or new and more sophisticated facilities to test their automotive products included Volvo of Sweden for an Increased Speed Upgrade of its Climatic Wind Tunnel, built earlier, and a new Model Aerodynamic Wind Tunnel built to 1/5 scale. Chrysler Corporation at Chelsea also upgraded the airspeed and control system of its wind tunnel in Chelsea, Michigan, and the General Motors, Rochester, N.Y., division authorized an engineering study for an Environmental Vehicle Test Laboratory. Their reputation as the designers and builders of wind tunnels not only spread to more European auto makers such as Daimler Benz and Behr of Germany, and

Saab in Sweden, it also won contracts from two major manufacturers in Asia, Hyundai in Korea and Honda of Japan.

Following his role of site manager for Porsche in Germany, Roland Vander Straeten promised his wife Rita that he would not take another overseas assignment unless it was "an adventure." A year later the chance to go to Korea as site manager for two climatic wind tunnels to test cars as well as buses and trucks occurred, and, he recalls with a smile, "I said to Rita, 'this is an adventure, as I have never been to that part of the world personally' and she agreed. So I went to Korea and then Japan to be site manager for the Honda wind tunnel." As it turned out, it was an adventure that led to the end of his career with DSMA International.

The first thing he did on arriving in Korea was enroll in a language course for two hours a day, five days a week for six months as he felt it was important to be able to be understand their culture and speak the language. "I know they appreciated that," he says, pointing out that he made friends with a number of company executives that proved beneficial when Aiolos Engineering was formed in 1994.

When the wind tunnel at the Hyundai Research Center at Ulsan, South Korea, was opened in 1989, a traditional ceremony - complete with the head of a pig adorning the festive table - was held. Al Bullock, then with DSMA International, was among those present to take part in the ceremony that also included drinking a glass of "Soju" at the event. Photo: Courtesy Al Bullock.

Shortly after setting up an office in Korea, DSMA signed a contract with Honda for an aeroacoustic wind tunnel in Japan. As Roland knew the manager of the Honda project he agreed to become site manager of it as well, on the understanding that he could obtain more help if or when it was required. He now had an office and an apartment at both sites and recalls, "I had two suitcases packed at each apartment and was continually flying from one place to the other. As a result, I had no spare time at all."

Besides the pressure of serving as project manager in two countries there were cost overruns on the Honda project and difficulties arose in solving them. Late in 1988, Roland flew to Toronto to ask Uldis Kiesners, his immediate supervisor, and Lorne Wilde, the vice president and chief operating officer, to let him have Alistair MacKenzie. "We had worked together at British Leyland and had a very good working relationship as well as being good friends, but he was tied up on another project. Gerrit was also busy elsewhere also, so I told them I would quit in four months time if I didn't get some help but I don't think they believed me." Four months later he was succeeded by Ratko Dzoja, a graduate of Zagreb University who worked as a miner at Flin Flon, Manitoba, when he first arrived in Canada in 1970. After several other jobs as an engineer, Ratko joined DSMA International in 1980 as a project engineer, spending more than two years in Sweden on the Volvo site. Later he returned to Yugoslavia to work on the tail end of the blowdown wind tunnel project that involved just about every senior person on the staff of DSMA International over a period of roughly eight years. It was the second turnkey project for blowdown wind tunnels signed with a government in Communist Europe, the first being one at Bucharest, Romania.

10 Success in Europe

An earthquake severely damaged a number of buildings during the construction of the blowdown wind tunnel in Romania. While the tunnel suffered only minor damage, some of the factories that built parts for the tunnel were severely damaged causing further delays in completing the project. A week after the earthquake, Doug Hunter arrived in Bucharest and took a number of photos of the damaged buildings in the downtown area.

An earthquake, added demands for the latest breakthroughs in wind tunnel construction, mistakes made by contractors, government bureaucracy and a severe economic downturn in the 1980s all contributed to the protracted time periods required to complete two blowdown wind tunnels built in Romania and Yugoslavia. Instead of the usual three to four years from contract signing to commissioning, they took six and eight years respectively. On the other hand, the fact that they were successfully built in countries behind the iron curtain encouraged both China and Russia to discuss similar projects with DSMA International when those two countries considered building wind tunnels. There were also other projects in Europe that further enhanced the reputation of the Canadian firm – projects such as a study done for NATO and the huge acoustic chamber completed in 1990 in The Netherlands.

Doug Hunter remembers the Romanian contract well. He was in India when he got a wire from George Meagher that Romania was interested in such a structure for its military establishment and suggested he fly home via Bucharest to find out more about it. He did, and, later, a delegation of Romanian officials visited the blowdown wind tunnel at Ottawa which led to further meetings in Bucharest. Doug remembers at one meeting the minister started by stating that if George was not prepared to reduce the price by 25 percent, why had he come? "George," Doug remembers, "pushed back his chair, closed his proposal book and said, 'if that is going to be your attitude, Mr. Minister, we are leaving.' The Minister backed down, and the meeting continued." Later, Hunter was a member of the team at the signing ceremony that took place on May 12, 1972, and photographed the group that included a number of Romanian government officials with George, Senator Daniel A. Lang, as a director of DSMA, and E. L. Bobinski, Canada's Commercial Counsellor from Vienna.

The contract called for the design and building, on a turnkey basis, a blowdown 1.2m X 1.2m trisonic wind tunnel for the Romanian Aeronautical Research and Design Institute. CPP was to carry out the turnkey project under the direction of Doug Hunter. The design called for a unit with wind velocities between Mach 0.2 to 3.5 and a test section

Signing a contract for the Romanian blowdown wind tunnel at Bucharest in May, 1972, are seated, left to right: Senator Daniel A. Lang, a director of DSMA , D. Atanasiu representing Tehnoimport of Romania, and George V. Meagher as a partner of DSMA. Behind are other Romanian government officials and E.L. Bobinski, standing far right, then the Canadian Commercial Counsellor in Vienna.

capable of testing three-dimensional models under subsonic and supersonic conditions. While much of the hardware was to be supplied by Canada, largely financed by the Export Development Corporation, the Romanian officials naturally wanted local companies to be used wherever possible, so Doug and Herb Kelland, who had been a key contract manager during the building of the blowdown wind tunnel at Ottawa, made a tour of various Romanian companies to inspect their capability for the project.

Some components built in Canada included the flexible nozzle at Davie Shipbuilding Ltd., Lauzon, Quebec, and the pressure regulating valve by Ebco Industries of Vancouver. The computer contract was awarded to Honeywell in Toronto. Two contracts awarded by CPP to companies outside Canada included the manufacture of optical glass ports in the observation room called a Schlieren system, to Optical Works Limited of England and parts of the variable diffuser and transonic section to NOHAB of Sweden. "It was our policy to look for the best and at the best price," says Doug.

By November 1974, Scottish-born Andrew Logie arrived in Bucharest to be project manager. On arrival he learned that because of housing shortages, the Romanian government had decided all foreigners were to be forced to live in hotels and those with homes were to be evicted from their residences. Two officials of the Romanian wind tunnel project, who had been among the dozens of Romanians sent to Canada to work on the project at CPP or attend training courses at the NRC in Ottawa, talked to the mayor of Bucharest. Andrew was allowed to live in an apartment with his daughter who was employed as his secretary.

Two years, however, became closer to four before the project was ready for commissioning by Bill Rainbird, part of the delay caused by an earthquake that severely damaged a number of buildings in Bucharest early in 1977. While the wind tunnel itself was only slightly damaged, some of the factories manufacturing components for it were literally destroyed adding significantly to the problem of completion, which finally occurred early in 1978.

By then, DSMA had completed another contract in Europe to further establish a worldwide reputation in wind tunnel expertise. Four NATO member countries - Great Britain, France, The Netherlands and West Germany - felt there was a need for a large transonic wind tunnel in Europe capable of testing aircraft models at conditions simulating the high Reynolds number found in actual flight. As it was felt that such a project was beyond the financial capability of any one country they formed a consortium and on submitting their various proposals, it was agreed the four schemes should be evaluated. After an international search NATO's Advisory Group for Aerospace Research and Development, dubbed AGARD, awarded the contract to DSMA, a project that was to place the company in a position to study the most advanced wind tunnel technology in the world.

The studies for AGARD involved many members of the DSMA Engineering staff, while others were deeply involved in the design and building of the Yugoslavian blowdown wind tunnel. Like the Romanian project, there were a number of reasons for taking even longer to complete (eight years) and involved just about every senior person at DSMA International before being commissioned in 1986. Among those who spent years at the site was Dobrilo Vucicevic, who was born and lived in Kraljevo, about 100 miles south of Belgrade, until coming to Canada in 1974. After stints with two other major consulting engineering firms in

Canada, he was hired by DSMA International in 1981 and after a few months in Toronto returned to Yugoslavia in charge of supervising the electrical and control systems.

Others included Gerrit Zorgdrager as project engineer, Gary Elfstrom and Steven Hills who, as a young engineer in his 20s at the time, was sent to help Gerrit for several weeks and ended up returning to Toronto two and a half years later, married to the daughter of the Brazilian consulate serving in Belgrade. He, along with Gerrit, and Dobrilo are members of the management team at Aiolos Engineering Inc., today. Another to serve in Belgrade for more than three years was John Farrell, now Associate Director of Planning and Construction Services at McMaster University, Hamilton.

The contract called for a blowdown unit with 1.5m X 1.5m test section size that could test three-dimensional models under subsonic, transonic and supersonic conditions. It was also to be equipped with two interchangeable test section inserts, the smaller one measuring 0.38m x 1.5m to be used to test two-dimensional models. The air was to be compressed in five storage tanks with a run time of between 6 seconds to 60 seconds in order to achieve velocities of Mach 0.2 to 4.0. "Flow quality in the test section is expected to be exceptional, conforming to the best of international standards," an early project sheet declared, but made no mention of some of the other features that were added over time. These included the introduction of splitter plates in the porous walls of the transonic test section and a new state of the art computer system developed by a newly established systems engineering group at DSMA International during the years the wind tunnel was being built.

Dobrilo, who was new to DSMA, remembers his fascination at how such a huge and complex unit had such tight tolerances in it. "I had never come across it in any place I had worked before," he claims. He cited, as an example, how the supersonic flexible nozzle overall structure was four to five metres high, three to four metres wide and 17 to 20 metres long yet the accuracy of the positioning of it was less than 1/10 of a millimetre. He also remembers the huge additional transonic test section that could be rolled into place on rails and, again, the accuracy allowed for only a couple of millimetres of tolerance.

Another factor in the time it took to complete the project was the additional features that were added to the original design. John Farrell, as the project's site manager, recalls the request by the Yugoslavian project

Among the major components designed and built in Canada for the Yugoslavian wind tunnel were the air storage tanks, made by Dominion Bridge, and a pressure regulating valve manufactured at Canadian Vickers. They were both then placed on a ship in Toronto harbour bound for Belgrade. Once there, the 50-foot air tank sections were assembled into five tanks. Photo: Courtesy John Farrell.

Dr. Gary Elfstrom checks the AGARD calibration model mounted on the rear sting support of the trisonic blowdown wind tunnel designed and built by DSMA International, near Belgrade, Yugoslavia. Numerous technological demands and an economic downturn were among the reasons the project took eight years before being commissioned in 1986.

manager that splitter plates be inserted in each hole of the perforated steel plates for the transonic test section in order to improve the acoustic quality of the wind tunnel. There had only been one other effort like it for a pilot wind tunnel in the United States but DSMA agreed to do it and a search began to find a company that could do the job. "There were literally thousands of holes in the steel plates, some 14,000 in all," John reports, "and eventually the Zurich office of DSMA International found a Swiss company that could make the splitter plates integral with each hole using the Electric Discharge Machining technique. They worked 24 hours a day for months to complete the job," John adds, "but that was DSMA International. When they said they would do something they did it."

Gary Elfstrom recalls the same request from a slightly different perspective, claiming the company's attitude was "Here was something new; so let's do it. But instead of costing a couple of hundred thousand dollars it probably cost a million and a half or more," he suggests, a factor that Richard Williams, as the comptroller of the company, found aggravating on many occasions. "We were continually underestimating engineering costs on projects, or letting the engineers develop concepts that overspent that portion of the budget," he claims "and that certainly happened on the Yugoslavian project."

There were other factors as well for the extended length of time it took to complete. A mistake by a local building contractor caused a gap of roughly ten feet in the wind tunnel itself "... so we had to make all sorts of compromises because we couldn't change the airflow of the tunnel itself," George Meagher remembers. "There was also a major downturn in the economy that caused all sorts of problems," he adds, "like power shortages so machines were shut down, or you couldn't get heat or petrol or parts. Actually our staff were quite ingenious in overcoming all sorts of problems like this."

Dobrilo and Gerrit who both spent more than four years in Belgrade, confirm George's observations "For a period they turned the electricity off on a rotating basis, Gerrit recalls, "from 6a.m. to 2p.m., then 2 to 10p.m. and 10 to 6 in the morning. I also remember my wife Alie waiting hours in line to get gasoline and by the time she got to the pump there was none left, and another time when we ran out of heating oil, and going to

Austria because we couldn't get any in Yugoslavia. I think the whole neighbourhood connected to my tank," he adds with a hearty laugh.

"But from a technical point of view we built a good wind tunnel there. It worked well. There were a number of innovative things done, such as the splitter plates," Gerrit adds. Gary Elfstrom attributes some of the problems to the fact that it was from a technical point of view, an extension of the art in certain areas." Sent to Belgrade for three to four months as the aerodynamicist to commission the wind tunnel, Gary stayed 33 months and took part in several of the problems he and his earlier colleagues faced. There was, for instance, a problem with the flexible nozzle contracted to a company that had never done one before. "They were a very good company mind you, but they used a rubber in the seals that was the same quality as the rubber used to make tires for their cars. Well it wasn't long before the seals were squirting oil everywhere and once a hydraulic system leaks you can't hold pressure anywhere. So the seals had to be stripped down and we found bits of crumbling rubber. Then we had to scramble around and go to Germany to get the proper rubber seals for the nozzle," he recalls.

He also remembers that the Yugoslavian wind tunnel was the first to be equipped with an integrated computer control and data system, one developed by what had become the systems engineering group of DSMA International. In 1979 George Meagher had hired Ron Lake, from Vibron, as a computer expert to head up the new division, and Lake quickly added a number of people to the staff including Arthur Ho as his assistant. Ho, a 1970 graduate of Hong Kong University in mechanical engineering who won a Commonwealth Scholarship to attend the University of Toronto, obtained a Master's degree in control systems engineering there. On joining DSMA International in 1980, he and other members of the

"The big woofer" is the name given to the unit used to create an overall sound pressure level of up to 154.5dBL. It is part of the Large European Acoustic Facility (LEAF) shown on page 131.

new systems group started a unique computer control, computer data acquisition and computer automation system and introduced it in the Yugoslavian wind tunnel. "Computers were very expensive in those days," Arthur admits, claiming that each computer system cost roughly $250,000 but had less power than today's PC. "Later on, as DSMA International was involved in several wind tunnel projects at the time,

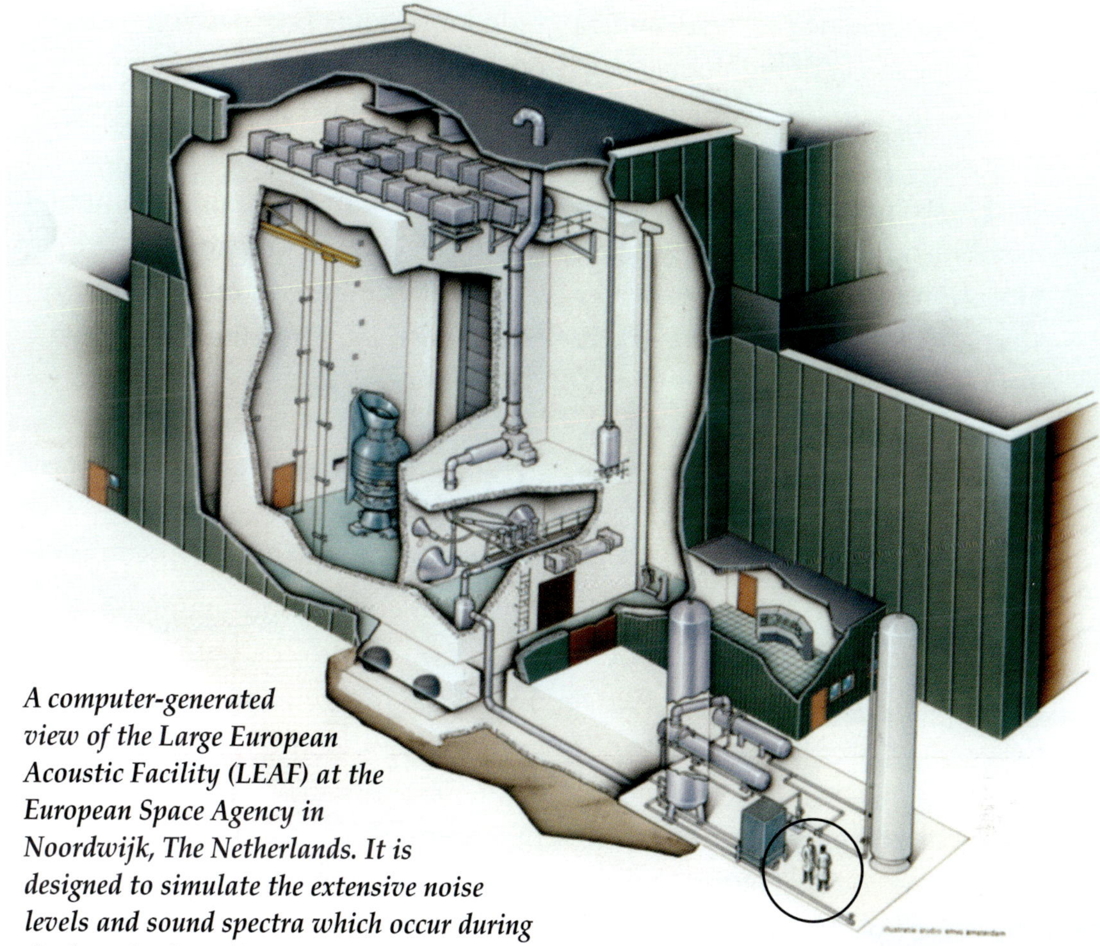

A computer-generated view of the Large European Acoustic Facility (LEAF) at the European Space Agency in Noordwijk, The Netherlands. It is designed to simulate the extensive noise levels and sound spectra which occur during the launch of satellites. An indication of the size of the project is reflected by the two figures shown at the bottom right of the drawing. Photo: Courtesy ESA.

we thought it made sense to standardize the software and we did that by filing a trademark using the name 'Talent' to identify the project. It was a great success, the only software system of its kind in the world," Arthur adds.

In 1987, there was a unique project known as the Large European Acoustic Facility (LEAF), a DSMA International turnkey project which was commissioned in 1990, and upgraded by Aiolos Engineering in 1996, to test the effect of high noise levels on spacecraft between the time of ignition and lift off. This happens because, an Aiolos Engineering project sheet stated "... one of the most severe environmental threats to a spacecraft is near the ground during the rocket launch. In the first ten minutes from lift-off, the space craft is subjected to structure-borne noise and vibration inside the payload bay," the sheet explained, and added, "Initially, the noise source is the rocket exhaust; later it is the high speed airflow over the payload bay."

To assess these concerns the European Space Technology & Engineering Centre (ESTEC), a division of the European Space Agency (ESA) in Noordwijk, The Netherlands, engaged DSMA International in 1987 to design and build LEAF. Gary Elfstrom, then chief aerodynamicist of the company, worked closely on the design of the project. with Walter Shickle of ESTEC and Frank Slingerland of the Institute for Aerospace Research of the NRC in Ottawa.

A structure large enough to test an entire Hermes spaceplane was called for, making the size of the chamber 11m long, 9m wide and 16.4m high. The noise spectra was to reach 154.5dBL, with a smaller structure, a mini-LEAF capable of achieving 162dBL, to be inserted inside the larger unit when required. The chamber had 0.5m thick steel reinforced concrete walls with a thick coating of epoxy paint applied to the inside surface to reduce acoustic absorption and improve the reverberation characteristics. The door, 7.6m high and the same thickness as the walls, alone weighed 800 tons and was moved on a wheel and rail system, and a 16-ton overhead crane with a hook height of 15m was designed to enable the test item to be suspended on cables from selected points on the ceiling and walls.

Once the design was approved, Alistair MacKenzie, on the staff of DSMA International since 1979, became project manager for LEAF spending more than 18 months on the site. "We employed a number of local contractors to build the concrete chamber, a mechanical contractor to build the massive door, and another local contractor to process the liquid nitrogen to gaseous nitrogen that was blown through some very large horns to produce the specified noise level."

Alistair said the horns were designed to generate noise levels up to 155 dBL. "The noise of a jet is about 150dBL, and that will cause hearing loss very quickly and 160dBL will actually kill you." While on site, he recalls watching from the ESTEC cafeteria a video feed of the launching of the joint ESA and NASA Hubble space craft from the rocket centre in Suriname, the former Dutch colony in South America. Alistair was also on site when Queen Juliana of The Netherlands attended the official opening with her son, who was studying to be an engineer. "The 80-ton door was opened by remote control because it wasn't working quite properly at the time and as it slid back, it revealed a satellite hanging from hooks in the chamber. It was very impressive," says Alistair.

11 Good Times and Bad for DSMA International

At Shenyang, the site where China's Research Institute planned to build a high speed blow-down wind tunnel, based on the one they had seen in Romania, DSMA and Chinese officials pose outside the schoolhouse where the meeting was held in late October 1980. Flanking the head of the Chinese delegation, Jiang Tung, are the four DSMA representatives, from left to right, Bob Stapells, George Meagher, Douglas Hunter and William Rainbird. Photo: Courtesy Douglas D. Hunter.

In 1979, China indicated an interest in building a wind tunnel similar to the Romanian project and George Meagher, along with Bill Rainbird and Uldis Kiesners, flew to Bejing to meet with Chinese officials. While Uldis remained in Bejing, George and Bill accepted an invitation to visit Mien Yang where the Chinese had a high-speed wind tunnel installation. They were the first westerners to visit the area in some 30 years and were stared at by the local citizenry, George recalls, while Bill noticed that the control valve was an actual copy of the one DSMA had installed in Romania. Nothing was decided on the initial visit but, later, George and his wife Evelyn entertained a Chinese delegation at their home in Toronto at which, George recalls, "They were all very orthodox in their Mao outfits and talking only in the presence of a third party interpreter."

The event, however, established a relationship with their leader, Jiang Tung. "When Evelyn and I visited China in March 1980, we were received with amazing hospitality and attentions." George claims, and remembers that one of the delegates who had been at their home suggested a painting they might like to purchase and even where it should be hung in their living room. In another visit to Meagher's home that summer, George also recalls, "The same Chinese delegates came dressed in western clothes and talked freely in good English, even commenting on Chinese political issues. It was of course after the displacement of the Gang of Four who had replaced Mao following his death in 1976," he adds. Even better, George recalled that it was during that visit that Jiang Tung told George that if his company came up with a good offer they would get the contract for the wind tunnel. In October 1980, a final trip was made to China, with Bill Rainbird, Doug Hunter and Bob Stapells as head of CPP, attending a meeting in a school house at Shenyang, the proposed site for the blowdown wind tunnel. There, a letter of intent outlining this agreement to George was delivered for his approval. With hardly a glance at the document, George signed it. After the Chinese official had left, Doug and others questioned why he hadn't even read it over. "No, if I had gone over that letter word for word, that fellow would have realized I didn't trust him and that would have been a bad beginning to the start of a relationship so I didn't take that chance," Doug recalls him saying. "That was George, " Doug adds.

In February 1981, however, China suffered a general economic collapse and George returned alone "... to join a small army of European businessmen waving cancelled contracts. I met Mr. Jiang and advised him we would seek no claims and wait for better times." It was a generous offer, given the trips, entertainment and preliminary designs had cost thousands but George was confident it was simply a matter of time before the contract would be confirmed. "By the late eighties, however, when China had recovered its economic strength and we could have exploited our highly favourable position there, circumstances and resources did not allow," George records in his account of the history of the company.

There were several reasons why "circumstances did not allow" for pursuing the contract. In 1982, George decided DSMA International would move into its own building at 10 Park Lawn Avenue instead of continuing in rented quarters with Lloyd's ATCON group at 4195 Dundas Steet. That was a mistake in the eyes of Richard Williams, a chartered accountant graduate who had joined the company as comptroller in 1968. There was inflation and double digit interest rates and the move was made largely on anticipated income based in large part on the signing of the contract for the turnkey blowdown wind tunnel in Yugoslavia, which ended up costing more than expected.

This, however, did not deter the ever-optimistic George. In March 1984, for instance, he wrote a glowing staff memo foreseeing, in the year ahead, "... somewhere between $20 to $25 million consolidated revenue, adding that in four to five years this might increase to a potential of $100 million per year or better." In the same memo he announced a number of organizational changes: George Bremner to become senior vice-president of the company as well as executive vice-president of Champlain Power Products and its two subsidiaries; R.F. Bob Stapells, vice-president of DSMA and executive vice-president of DSMA-Tata, the latter based on the belief that DSMA-Tata was anticipating a major increase in its nuclear and other India-based activity and because Canada was about to re-instate loans and aid to India; and Uldis Kiesners, appointed vice-president and general manager for Champlain Power Products to administer all contracts undertaken by the company.

George's bright future, did look promising in 1984, with more than $19 million realized in new contracts, largely due to a contract with Porsche of Germany, but the following year dropped to $11.8 million and in 1986 the total fell to just over $7.5 million. With the continued downturn of the economy and growing losses in overall company operations, at least three of the outside directors, Michael Koerner, Senator Daniel A. Lang and Lorne Webster expressed their concern and later resigned. In 1987, the Park Lawn building was sold "at a profit of nearly three million dollars" George quickly points out, "to reduce company debts." The company then moved to rented premises at 6655 Airport Road. By then George had become chairman and CEO, George Bremner was appointed president and Lorne Wilde, an 18 year veteran of the company, named vice president and chief operating officer.

Despite the downturn, George continued to be optimistic and believed the future for contracts was sound. On a trade mission to Spain with

Finance Minister Michael Wilson, for instance, George believed the only sale the Minister would come home with was a DSMA wind tunnel project if the Minister pushed it, suggesting at one meeting, "Look, Mr. Minister, "you'll have something to take home from here, the mission will have some measure of success, and I'll have $10 million dollars. But it didn't happen," he adds.

In 1988, there was a glimmer of hope when Sandia National Laboratories, a research arm for the United States Department of Energy, wanted a feasibility study on a wind tunnel to test parachutes that were used to deploy cargo and/or weapons at a low altitude. The parachutes were collapsing and no existing wind tunnel could simulate the air flow required so they asked DSMA International to determine what if any kind of facility they would need. Gary Elfstrom, who early in his career taught aeronautics at the University of Tennessee, recalls, "It was a fun project, something really far out. We looked at several options with David Van Every doing most of the aerodynamics for it, right down to the basic study on the validity of simulation in a wind tunnel. We finally came up with a concept involving a projectile model travelling through a long slotted wall test section. Sandia was keen to give us a contract but the military who were to be the sponsors thought $30 to $50 million to solve a single problem was a lot of money. But it gave us a fascinating glimpse into how far ahead the U.S. military is in R &D. When we raised a projectile survivability problem, for example, they saw it as no problem because they already had a little system, not much bigger than the end of your thumb, that they can shoot out of cannons, thousands of them, and then recover them, plug them into a computer and record the data."

That loss, however, was offset in 1988 when DSMA International signed a turnkey contract with Honda Corporation of Japan for an aerodynamic wind tunnel that, with other contracts that year, amounted to total sales of nearly $48 million, the biggest single year in the company's history. Besides the huge $34 million contract with Honda, there was an agreement of some $3.5 million signed with a company in Finland who were commissioned to build an automotive aerodynamic wind tunnel like the one at Porsche and a climatic wind tunnel similar to that of SAAB as well as other climatic chambers for the huge Volga Auto works (VAZ) that manufactured Lada vehicles at Togliatti in the USSR.

Ratko Dzoja recalls the visit of the Finnish company EKE representative. "I was working on the design of the Honda project at the time when George Bremner came to my office and said he had a chore for me. He wanted me to come out and see this man who has $300,000 as a down payment for us to design a wind tunnel for Russia. They had seen the one at Porsche in Germany and wanted one like it."

A proposal for the detailed design engineering for the project was quickly submitted. It began by pointing out that since its founding in 1978, DSMA International had executed more than 150 projects in 15 countries. It claimed that its slotted wall designs for automotive wind tunnels "developed from an intensive in-house R&D program funded by DSMA International, represents the leading technology for automotive aerodynamics testing today." It explained that the slotted wall technology provides significant economies in nozzle and return circuit sizing while maintaining very accurate and reliable airflow and surface pressure simulations, citing its Volvo of Sweden and Porsche of Germany as "successful slotted wall projects." It also stated that since the first wind tunnel designed for Chrysler, more than 50 automotive facility projects were either completed or in progress.

It highlighted numerous other accomplishments such as the proper selection of nozzle size for climatic wind tunnels that has "a first order of impact on the extent of the vehicle (front, rear, underside, interior etc.), which receives proper airflow simulation." It recommended its new adjustable collector flap system that had been used in its two most recent climatic wind tunnel designs for Saab of Sweden and Ford U.S.A. It also suggested its use of the "Idle/City Traffic Simulation System," successfully used in 11 completed facility designs, and reported on how DSMA had "... conducted a research program to identify and incorporate lamps having a more faithful replication of the spectral energy contest of natural sunlight," and had recently come up with the use of a Dutch quartz halogen dichroic mirror lamp that had been used by GM-Buick, Ford of Great Britain, and again, Saab.

While the company admitted it did not manufacture dynamometers, it claimed credit for a unique sliding slot cover system in answer to the troublesome dynamometer/floor design interface. "To date, this sliding slot cover system has been incorporated in facility designs for GM-Buick, Ford U.S.A., Ford of Great Britain and Hyundai in Korea," and a final feature, highlighted in the proposal, dealt with the heat loads generated by radiation from the road surface and increased underhood and underbody temperatures because of "tighter packaging and increased shrouding of vehicle underbody components."

DSMA won the contract that eventually amounted to roughly $3.5 million. It included detailed drawings based on the Porsche project "... so it wasn't difficult from a design point of view," Ratko recalls."But working with the Finns and the Russians was different," he adds. "They were very stage conscious, certain things had to be done in stage one, and then stage two and three, and we were not used to that." In 1989, Ratko was

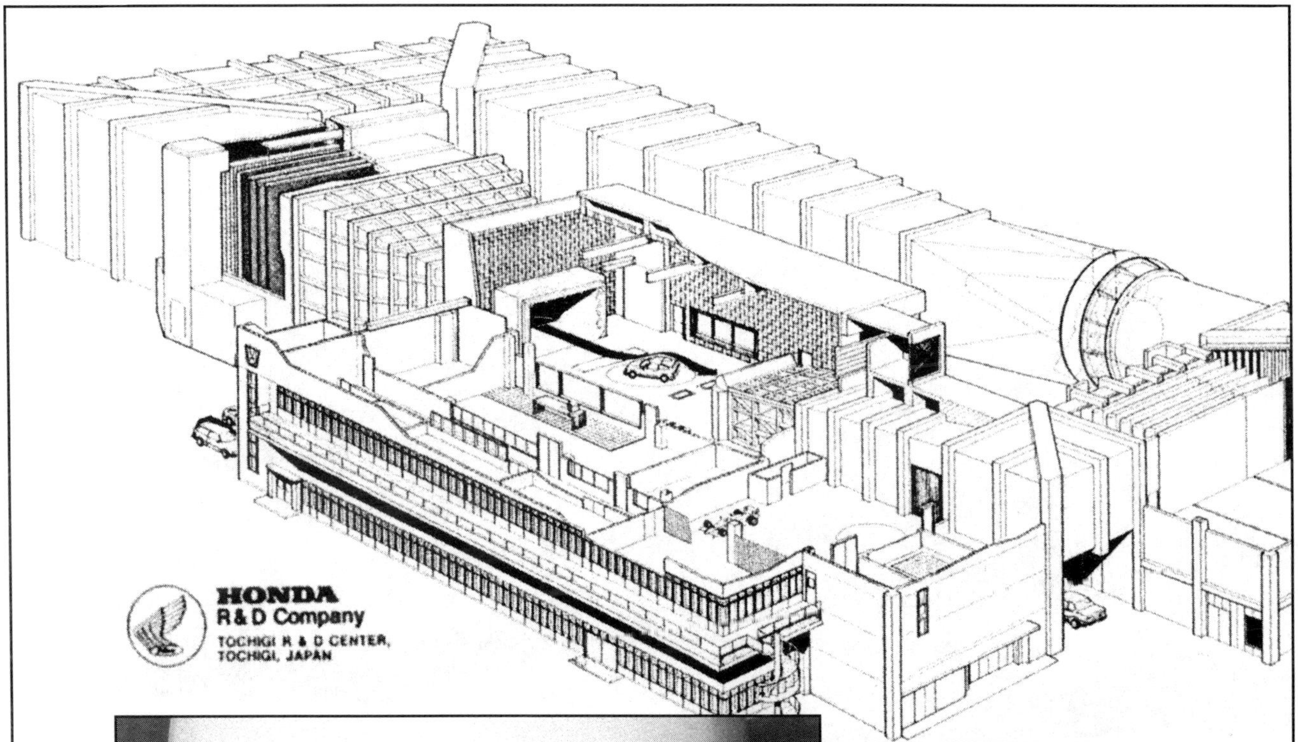

During the commissioning of the huge aeroacoustic low speed wind tunnel at the Honda Research and Development Centre at Tochhigi, Japan, Peter Waudby-Smith and David Van Every take a moment to indicate the size of the fan David designed for the project that was opened in 1991. The fan is 9.2m in diameter and can produce winds in excess of 200 km/h. Photo: Courtesy Honda R & D. Company.

off to Japan to replace Roland Vander Straeten and stayed there until the end of the project. While in Japan, Ratko was appointed a vice-president of DSMA International by George Bremner, who became president in 1987, and Lorne Wilde was appointed senior vice-president and chief operating officer. In 1990, Graham Thomas went to Finland to complete a scale model of the wind tunnel for VAZ.

By then, however, issues other than securing contracts were surfacing at DSMA International. The Honda contract had cost overruns amounting to some $8 million "because of a low estimate initially" in the opinion of Ratko (a deficit later recovered from Honda), and the only job on hand was one for less than $5 million with Williams Grand Prix Engineering in England. Nevertheless George soldiered on, believing he could still bring in a project that would put everything back on its feet, but for DSMA International it was too late to stand alone any longer.

When nothing in the way of contracts materialized in 1990, some key members of the staff obviously began wondering what the future held for them. Arthur Ho, who headed the systems engineering group and its computer software, known as Talent, recalls that he, along with Lorne Wilde, Gary Elfstrom, Joe Templin, Ratko Dzoja (home from Japan after the company successfully renegotiated its contract with Honda) and others would lunch together and consider their options. "There was no conspiracy with anyone else about this," says Lorne, and Ho agrees. "We just met and discussed where the company was going and what were our options." Ho was the first to act when he, along with his assistant Peter Tong and two other members of the group, made a deal with George to leave DSMA International and form ReACT, a company that is now a leading supplier of sophisticated computer software for industry throughout the world. "We left for several reasons," he explains. "We found that being part of a mechanical and electrical consulting firm, we couldn't develop the kind of things we wanted to do in computer technology." After the sale of DSMA International to TLT Babcock, Ho also made a deal to purchase not only the Talent trademark, but also the patents, other intellectual properties, software and equipment that had belonged to the former systems engineering group.

In June 1991, Lorne Wilde, a small shareholder and one of three staff members who had signed a personal guarantee on a DSMA International loan at the bank, demanded changes and when they weren't forthcoming, quit without a job in sight. He was followed shortly afterward by Gary Elfstrom, Joe Templin, and Gerrit Zorgdrager before the end of the year. Dick Williams, the comptroller who, along with George Bremner, had also signed a personal guarantee for the bank loan on behalf of the company, also left to open his own office as a Chartered Accountant.

Meanwhile, George Bremner, as president, tried to keep the company afloat by pursuing several objectives. He entered an agreement with Mitsui Corporation of Japan to sell them some of DSMA's intellectual properties. He also flirted with TLT Babcock of Akron, Ohio, a fan company with limited experience in the wind tunnel business but interested in expanding into that field. When that came to naught and a group of former executives of Frank Stronach's Magna Corporation showed an interest in buying the firm, George Meagher, as chairman, resumed day to day control of the company and appointed one of that group, Don Parker, to take over as president. The situation lasted only months, however, as the group voted against purchasing the company, and Don Parker quit.

George Meagher then pursued a more alluring option: the design of a huge low speed wind tunnel for Boeing Aircraft of Seattle. "I established a joint venture undertaking with Deutsche Babcock of Germany and carried on the negotiation with Boeing for a turnkey contract for the world's largest low speed wind tunnel. It had a budget in excess of $100,000,000 U.S.," he adds, "and was a very arduous negotiation." It included, as a corollary undertaking, that TLT Babcock, the American subsidiary of Deutsche Babcock at Akron, Ohio, would buy out his share interest, take over DSMA management and "leave me in an advisory role of semi-retirement," George noted. "With the company apparently salvaged, I entered hospital for much overdue bypass surgery," he adds.

On signing this agreement, Walter Beller, president of TLT Babcock, realizing they had little experience in wind tunnel projects, hired Lorne Wilde and Joe Templin for the contemplated Boeing project. Beller added Gary Elfstrom to work out of the Toronto office while most of the staff at DSMA International – now numbering some 70 employees – also felt relieved as they began preliminary designs on the Boeing low speed wind tunnel. On February 14, 1992, however, Boeing, the victim of cancelled aircraft contracts, created a domino effect by cancelling its plans to go ahead with the project. TLT Babcock then made a new proposal to George Meagher: that he sign a statement of bankruptcy to enable them to buy DSMA assets without assuming its debts. This greatly upset Fred Bauer, the DSMA representative in Switzerland who had negotiated many of the contracts and sub-contracts for the Williams wind tunnel in Great Britain, as the sub-contractors would not be paid. "One contractor, in fact, went bankrupt because of this," Fred claims.

The agreement also meant reducing staff without penalty, and TLT Babcock then set up a new subsidiary to be called DSMA-Babcock with Ratko named general manager of the group that was to remain in

Toronto. "It meant letting a lot of people go," Ratko recalls, "as the Toronto office of DSMA Babcock only had enough work for 10 or 12 people. That was a nightmare – the worst experience of my life, and it didn't get any better when at the end of that month I had to tell the remaining staff that they weren't getting paid because the bank had seized all the assets of the company," he recalled when interviewed at Toronto in the summer of 2000, while visiting the city from his current job as site manager for an Aiolos Engineering turnkey contract for the Universiti Teknologi in Malaysia. He is now site manager for the RTR wind tunnel in Vienna.

Ratko also recalled that DSMA-Babcock opened an office near the airport in Toronto until 1993 when a new president, Arnfred Kulenkampff, replaced Walter Beller as president at TLT Babcock and decided that DSMA-Babcock "... wasn't making enough money in the wind tunnel business, and put it up for sale." He also moved Lorne Wilde and Joe Templin off wind tunnel activities to develop test equipment for diesel engines that he was interested in promoting in the United States. Before the year was out, however, there was so little interest by U.S. buyers that he also shut that project down and suggested both seek other employment. Lorne became an executive of Sverdrup, DSMA International's major competitor for worldwide wind tunnel business, while Joe returned to Canada to become head of the aerodynamics laboratory of the NRC in Ottawa. Two other former staff members of DSMA remained with TLT Babcock in Akron, Ohio – Gerrit Zorgdrager, who, in 1992, became chief engineer of TLT Babcock's fan division and Graham Thomas, a young mechanical engineer hired by Gerrit at both DSMA International and TLT Babcock. In time, however, both returned to Canada to work for Aiolos, Gerrit now vice-president and general manager, and Graham now engineering manager.

Meanwhile, a scramble ensued for the purchase of DSMA-Babcock. The newly created, in name only, Aiolos Corporation, led by Roland Vander Straeten and Dobrilo Vucicevic, sought an opportunity to bid on the company. Ratko, with possible contracts from Ford Motor Company of Dearborn, Michigan, and a potential one in Italy, considered ways and means of competing. Another company, Fluidyne, was interested but eventually Ransco of Oxnard, California, manufacturers of, among other things, calorimeters for testing automobile radiators, bought the company from TLT Babcock and decided to close the Toronto office. In doing so, they scrapped literally thousands of drawings and other material from the Toronto office which Roy Fedoruk rescued from the recycle bins at the back of the building and stored in his garage "in case they came in handy some day." They also offered the Toronto staff an opportunity to move to their headquarters at Oxnard, California.

The terms were most generous. The staff were invited to visit Oxnard but several things worked against Ransco successfully attracting them. Adrian Hirsch made the trip but he and his family decided they didn't want to leave friends and family now living in Toronto. Several others felt the same way, while some were reluctant because that area of California had just experienced an earthquake with the damage still obvious when the visits took place. It was only days after the earthquake, for instance, when Mike Higgins and his wife arrived in Oxnard and saw pieces of road missing and debris laying about. "That didn't alarm us until that night when we had a room in a hotel on the beach and the aftershocks made us rush out of the building faster than you could blink," Mike recalls. Despite that experience Mike, a native of Blackpool, who, after ten years working for the British Aircraft Corporation as an aeronautical technologist, came to Canada and joined DSMA in 1976, decided to go because it appealed to him and his wife on several counts.

"I really didn't have another job to go to, we wanted to get rid of the house, my wife was able to take a year's leave of absence from her job and we thought a year in California would be interesting," he remembers.

Only one other employee made the move – Michael Doiron – but a month after being in California he was sent to Ransco's Detroit office and now works in Connecticut. Mike Higgins remained at Oxnard for 11 months, by which time, "... the company realized they were not really equipped to be in the wind tunnel business. Certainly just having me added to their staff wasn't enough," he adds with a laugh. "They really needed to flesh out all the disciplines that they lacked and realized they just weren't getting anywhere so I came back and joined Aiolos." He works there now as Head of the Mechanical Design Group.

DSMA as a title died with the purchase of the name by Ransco, but to many of the veteran members of the staff, it still conjures up a company of incredible talent and great challenges in the field of consulting engineering. Doug Hunter smiles warmly at the thought that he would never have seen much of the world if he hadn't accepted George Meagher's invitation to join the firm in 1963. "I really enjoyed working for DSMA. It was a unique company. No other company in Canada was as versatile and capable of designing things as well as DSMA, and as far as wind tunnels are concerned, there is no other company in the world that had their expertise and capability." In his opinion, the banks were too cautious or didn't understand the nature of a consulting firm. "Here was a company doing well, well-respected around the world for their expertise, but when some financial problems occurred, they just wouldn't wait, they wanted their money." On a personal note, he adds, "I toured the world, practically – India, China. I certainly had experiences I wouldn't have had anywhere else."

John Farrell, when interviewed, began, "My emotional attachment to DSMA is still alive today. We became the busiest firm in the world designing astronomical telescopes. In all our projects we had an engineering attitude to produce the best facilities in the world, and when DSMA said they'd do something, they did it."

Peter Tillson describes working at DSMA as "an interesting team-work approach. It was a systems approach where you shared the difficulties with others – the control problems, the mechanical problems, the stress factors and so on – so you were involved in a widespread series of engineering fields. That's what made the work so interesting."

Bill Rainbird, who split his Canadian career as a professor at Carleton University and as a DSMA executive between 1971 and 1985, calls DSMA "... a Canadian success story. The close co-operation between NRC and DSMA was important, and the team put together by DSMA in the late 1970s was about as good as you could get. We tackled some mighty big projects very successfully and the company was close to my heart for it was a major challenge for all of us."

Another to praise its technical skills was Michael Koerner, an engineering graduate who served as a director for several years. He claims the company was "a highly creative consulting firm with some really remarkable people." But as president of Canada Overseas Investments Limited and an expert in financial management, he once made the observation to Paul Dilworth, while he was still a director, that it was a unique experience to be a director of a firm with a negative net worth on its balance sheet. In the late 1980s he resigned from the board but still claims: "Technologically, it was a company Canada should be proud of, for it became a world-recognized firm in wind tunnel research and a clear leader in wind tunnels globally."

George shakes off the financial observations with a shrug. "We just had too big a collection of problems. We were overstaffed, I'll admit that, and contracts we had lined up got deferred or cancelled because of the economy. On the other hand, we persuaded a lot of the world that Canada was a good place to come for advanced technology and I think the proof of the pudding, in this regard, is that each of our projects was successful technically." He also sites the Hindu philosophy which believes one does not truly die but re-emerges in a greater or lesser form, adding, "There are several companies that sustain their life in some degree from the roots of the DSMA tree." ReACT and Aiolos Engineering are his Canadian examples.

12 The Founding of Aiolos Corporation

April 3/92

FOR NAME OF CO

393 University Av. 2ND Floor (Just S. of Dundas)
Consumers & Commercial
ph. 593 - 8880

1) AEOLUS - english speling
(ē'ə les)

2) AIOLOS - L, fr. Gk

3) E OLOS - how Greeks pronounce it

Alpha	α	A
Jota	ι	I
Omicron	ο	O
Lambda	λ	L
Omicron	ο	O
Sigma	Σ	S

There are 3 companies registered under AEOLUS:

AEOLUS Contracting Ltd.	Ontario	8/12/83
AEOLUS Investment Limited	"	29/05/73
AEOLUS Laboratories Limited	"	9/4/76

but no Co. registered under AIOLOS nor EOLOS !!

Let it be **AIOLOS Corp.**

This name has to be checked in USA and then registered. (NUANA ?)

Dobrilo Vucicevic volunteered to come up with a company name and after careful research, took his hand-prepared 8.5x11-inch document to the Ontario Government to register the name on April 3, 1992.

On April 3, 1992, Dobrilo Vucicevic registered with the Ontario government the name Aiolos Corporation with the names of four people identified as the participants: Roland Vander Straeten, Carl Balkwill, Sergio Raimondo and himself. It was the outcome of "chit-chatting," in Roland's car on a trip home from Ottawa when the four of them talked about an appropriate name for what was then merely an idea of forming a company. Dobrilo suggested he be the one to come up with the name.

The basis for choosing "Aiolos" was quite simple and appropriate. Aiolos was the "keeper of the winds in Greek mythology" who lived on a floating island – believed to be one of the Aeolian Islands near Sicily. The legend is that Zeus entrusted Aiolos with the power over all winds, which were stored in a leather pouch in a cave, and that Aiolos, after hosting a friendly visit with the Greek hero Odysseus (Ulysses in Latin) gave him the bag containing the winds to blow his ships safely home to Ithaca. Odysseus, however, fell asleep and his sailors, suspecting that the bag contained treasure (wine, in another version of the legend), opened the bag, letting the wind escape to cause a violent storm that drove the ships back to the Aeolian Islands. Aiolos, feeling that Odysseus was "the victim of devine wrath, refused to have anything more to do with him and sent him packing," according to one English version of the legend.

Regardless of legend, Dobrilo liked the name for all the reasons he had in mind. It met his wish to be short, catchy and memorable. At first, however, his colleagues were dubious and had to think about it. When Roland asked his wife Rita, had she ever heard of Aiolos, she promptly said it was obvious he didn't do crossword puzzles, and another colleague discovered that when he tried it out on friends, they questioned what it meant, discussed it and then "never forgot it afterwards."

Agreeing on a name, however, proved to be the easiest part of establishing Aiolos Corporation. For nearly 20 months, while Roland and Dobrilo were both employed with other companies, they spent most of their other waking hours attempting to bid on contracts or gain financial backers for the company. They were faced with the fact that, as a loose gathering of people in an under-financed organization with only the home address of either Roland or Dobrilo, they weren't taken seriously when making serious bids on wind tunnel contracts. When they tried to get financial backing they were equally stymied because potential backers wanted to know what contracts were on hand or at least in sight.

Their first plunge into bidding on a wind tunnel and procuring financial support came just two months after obtaining the registration of their name as Aiolos Corporation. Aware that Ford Motor Company of Dearborn, Michigan, were considering a wind tunnel, the newly named Aiolos company got the Cole Sherman Company of Toronto to write to Ford to confirm that the Cole Sherman company was interested. "We are endeavouring to finalize forming a business relationship with Aiolos in the very near future allowing us to jointly participate in any form of proposal submission to the Ford Motor Company or other potential clients."

Nor did they limit their search for finances at that. Days later Roland, Carl Balkwill, Sergio and Alistair MacKenzie met Loren Klevering, Group vice-president of Giffels International in Detroit, to exchange ideas on how the two firms might work together. Klevering made no commitment but asked that they provide resumes of the people associated with the new Aiolos group. Roughly 15 resumes were submitted but little else materialized. Two other firms in Detroit were also contacted: Albert Kahn Associates and ENGA, a major facility design and supply company, but nothing materialized with these contacts either. A similar fate occurred as a result with Acres International Limited in Toronto and Hatch Associates Limited, also in Toronto.

A meeting was also held with Asea Brown Boveri Incorporated (ABB) in Ottawa and, while ABB were not interested in joining forces, they did offer Sergio Raimondo a job in their engineering department and appointed Roland their marketing manager to the Pacific rim countries. Sergio moved to Ottawa while Roland was able to work out of his home in Toronto when required. Dobrilo also accepted a job as the Canadian marketing representative of a U.S. firm, and as such was also able to work out of his home in Mississauga.

But their jobs didn't stop them from continuing the search for financial backers and contracts that would lead to the actual establishment of Aiolos. They wrote to the president of Ransco Company of California, Jim Wasson, briefly outlining the skills of the Aiolos group and mentioned the Ford Motor Company as a potential opportunity, which prompted a meeting on July 8, 1992. Ransco stated they would not be prime on a turnkey project as $20 million was too large a sum and that they would only give financial support "... if Ford guarantees the project to Aiolos and if Ford underwrites Aiolos for five to six months."

It was another setback in a continuing series of setbacks. By then, however, this turndown was somewhat academic because Harvey Griggs, chairman of H.G. Engineering whose own offices were only blocks away from the old DSMA rented premises on Airport Road, had become interested. By a twist of fate, the man who brought Roland and Griggs together was George V. Meagher, ever the optimist, hoping to re-establish DSMA in some form. He had been in touch with Roland over that summer and had approached Harvey to see if he was interested.

Harvey, a structural engineering graduate of the University of Toronto in 1961, with a Ph.D. from the Massachusetts Institute of Technology, in 1969, had opened his own company, H.G. Engineering, in 1973, but had long admired the work of DSMA as an engineering consulting firm. He was impressed that they had turned the first wind tunnel at Ottawa into worldwide contracts, that they had created another worldwide reputation in telescopes after their participation in the Queen Elizabeth II project and that their nuclear and fusion activities were world class. He was well aware of their accomplishments because his own consulting company had done some stress analysis work for DSMA ATCON before it folded in 1986, "... so I considered this a wonderful opportunity."

Following a meeting with George Meagher, Harvey had a second meeting with Roland and the other members of the Aiolos team, Carl, Sergio and Dobrilo, and decided to recommend to his own board of directors that they back the group. They agreed, and a move was made to see if they could buy DSMA Babcock and the DSMA name. That, however, proved too expensive as it entailed taking over a five-year lease on the property then being rented by DSMA Babcock in Toronto. Without the purchase, they could not get the DSMA name.

"So, we had a choice," Harvey recalls. "We could abandon the project or we could start the company under the name Aiolos, which we all found difficult to pronounce at first but found easy later on." The latter choice was made in January 1994. Harvey squeezed the first employees into a corner of his own office space, provided a conference room when required and launched them on a "shoestring budget."

Their first project was also a "shoestring" one. Several years earlier, DSMA International had built a small wind tunnel for Frank Stronach's

Magna International Inc., in Markham, Ontario, to test radiators. This unit was later sold to CH Auto Parts Company of Thailand, where it was kept wrapped in a canvas for a few years. The company decided to reactivate it, and were interested in having the original designers of the unit carry out the work. They arranged to have Ziggy Polanski, of Magna International, come to Thailand on a leave of absence from Magna, and he, in turn, wanted Carl Balkwill, who had designed the controls for the wind tunnel to join him for a $25,000 fee. Aiolos agreed and sent Carl to Phrapradaeng, Thailand, where Ziggy, on behalf of CH Auto, was equally interested in seeing that the project was done on a shoestring budget. Ziggy suggested, therefore, that Carl could stay at his home as his family were away and could drive to work with him daily so he wouldn't need to rent an automobile. Carl agreed and enjoyed his stay in Thailand for the next three months as the wind tunnel was reassembled and tested.

That was the first contract to get Aiolos started. Shortly afterwards, Roland was a member of a group of Ontario businessmen who attended a trade mission, sponsored by the provincial government, where he met Glen Schuyler of RWDI, a boundary layer wind tunnel company in Guelph, Ontario, that ultimately resulted in a contract to design a damper needed for the four corners of a bridge walkway between the twin towers of 101 storeys of the Petronas building in the City Centre of Kuala Lumpur, Malaysia. Three dampers were installed in each of the four corners of the walkway.

Aiolos was then invited to have four people participate with several other American companies in a major National Wind Tunnel Complex project in the United States being carried out by NASA and Boeing Aircraft. Gary Elfstrom, Roy Fedoruk, Sergio Raimondo and David Van Every were assigned to the multi-consultant, kick-off team and worked

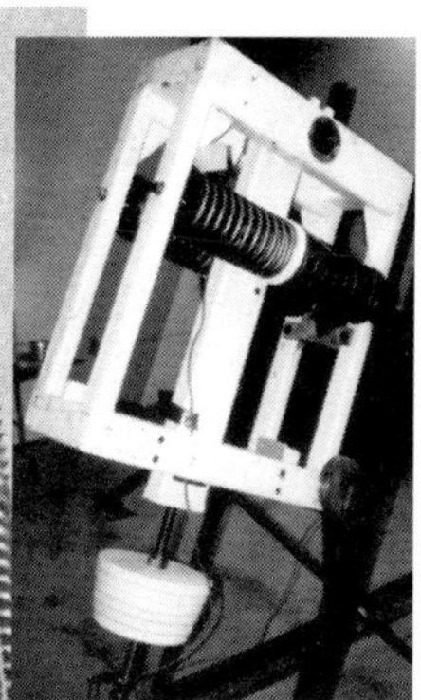

Tuned mass dampers are one of the cost effective means of suppressing excessive vibrations caused by wind on tall buildings. One of the first structures to be so equipped was the 101-storey Petronas Building in Kuala Lumpur. The damper (inset) was designed by Aiolos Engineering shortly after the company got started in 1994, and three were installed in each of the four corners of the bridge joining the two buildings. Photo: Courtesy RWDI Inc.

out of Boeing offices in Seattle, until a larger team took over in a NASA office in Cleveland. Later, all four received "Pride of Excellence Certificates" from the senior management of the project and a covering letter to Roland as president of Aiolos stating that "Their knowledge, skills and attitude are among the best we have encountered and we want to make sure that their significant contributions are recognized."

While the four were on that assignment, other projects began to come in as news that Aiolos was now in the wind tunnel business, with many former staff members of DSMA International making up the new firm. There was an engineering services study for a driveability test facility at

Ford Motor Company in Michigan and a study for an articulated sting system for the blowdown wind tunnel in Ottawa. Then Aiolos joined a German company, Turbo Lufttechnik, for preliminary design and engineering services for two wind tunnels for the Centro Italiano Ricerche Aerospaziali (CIRA), near Naples, which would amount to some 30,000 hours of design work. This enabled Aiolos to welcome back more ex-DSMA International technical people and the return of Dobrilo Vucicevic in a management role. By the end of 1994, a dozen former employees of DSMA International and ATCON were on the Aiolos payroll. Besides Roland and Dobrilo, there were Carl Balkwill, Al Bullock, Roy Fedoruk, Adrian Hirsch, Barry Jack, Dan Simec, Sergio Raimondo, Peter Waudby-Smith, David Van Every and Gary Elfstrom.

The CIRA project turned out to be a four-year involvement as CIRA itself gave Aiolos a contract to provide engineering services for an electrical control system for its Plasma Wind Tunnel that was completed in March 1995, and another engineering services contract for an insert system in its transonic model wind tunnel, completed in July 1995. These were in addition to the first contract with Turbo Lufttechnik which called for engineering services of a low speed wind tunnel, with a slotted wall /open jet test section measuring 7.2m X 5.5m and a second 4.4m X 5.5m solid wall/open jet test section. A second contract with the same German company for CIRA called for an Icing Wind Tunnel capable of having three test sections of the solid/slotted wall configurations and a fourth with a test section measuring 2.25m by 2.35m open jet test section. At least ten projects were either completed or underway by the end of the first full year so that Aiolos was able to not only show a profit but also pay back the $30,000 start-up fund to H.G Engineering. Now, seven years later, Aiolos has recaptured, for Canada, its earlier image as one of the major designers and builders of wind tunnels around the world.

13 Canada's Role in Wind Tunnel Testing

Professor Alan Davenport, Ph.D., Director, University of Western Ontario Boundary Layer Wind Tunnel, kneels behind some of the numerous building models from around the world that have been tested at the university since he developed the first wind tunnel there in the mid-1960s as a branch of the Faculty of Engineering Science. Bridges are another major aspect of the wind engineering studies carried out at the university located in London, Ontario. Photo: Courtesy University of Western Ontario.

There are three establishments in Canada that have an outstanding record when it comes to making wind tunnel assessments - the National Research Council of Canada (NRC) for aerodynamic testing, the University of Western Ontario and a private company, Rowan, Williams, Davies and Irwin, better known as RWDI Inc. – the last two for boundary layer wind engineering studies. Between the three of them, they have built an enviable record of wind tunnel testing worldwide.

The pioneer establishment for such testing was the National Aeronautical Establishment, now called the Institute of Aerospace Research (IAR) of the National Research Council in Ottawa. Building its first tunnel in the mid-1930s, it has, today, four constantly active wind tunnels at two Ottawa locations carrying out not only state-of-the-art aerodynamic assessments, but also doing research of its own as well as research for other government departments, agencies or private industry.

Aiolos Engineering, for example, currently has a study being carried out at the J.B. Parkin Building on Montreal Road to explore new effects of open jet wind tunnel configurations. It is under the direction of David Van Every, a University of Waterloo B.A. Sc. (1980), and M. Eng., in Mechanical and Aeronautical Engineering from Carleton University in 1983, who joined DSMA in 1984, and is now senior aerodynamicist and acoustician at Aiolos. One study that was initiated in 2000 with Kevin Cooper, the principal research officer of low speed aerodynamics of NRC, is testing wind tunnel configurations to better determine how to correct boundary interference using a computational model developed by NRC.

"Another is a study of slotted wall technology for racing cars," David reports. "We developed slotted walls for the automobile industry and it has been very successful, but now we're involved with racing cars whose designers have always wanted their vehicles tested in solid wall wind tunnels. We want to prove these vehicles could be tested with possibly better results and at less expense, so we are pursuing, with NRC, an investigation of the slotted wall test section with race car type configurations," David adds.

In working with Kevin Cooper, David has exposure to the expertise of one who has, since graduating with an M.A. Sc., from the University of Toronto's Institute for Aerospace Studies in 1968, spent his entire career at NRC specializing in research on the dynamics and aerodynamics of both surface vehicles and ground-based structures. Kevin's research has included thousands of hours on heavy-duty truck aerodynamic development, as well as the aerodynamics of racing ranging from skidoos and motorcycles to Indy cars. He has also spent time with numerous sports organizations studying the effects of aerodynamics for skiers and cyclists bent on winning gold at national and international events.

The author of numerous papers and articles, Kevin claims the use of wind tunnels has become a very big business. "Most automobile companies," he says, "have tunnels that run two and a half shifts a day testing every model of car many times." He believes cars are, in an aerodynamic sense, now reaching their peak efficiency with respect to drag which is created not only by the wind pushing against the front but the suction created as it passes over the rear of the vehicle. "To go further would mean having to maybe compromise the whole utility of the vehicle a bit, closing in the rear end more, for example," he claims. "I think the biggest efficiency, besides the computerized engine controls, has been drag reduction. When the drag coefficient goes from 0.5 on average in the 1960 vehicles to 0.3 to 0.4 on present-day vehicles, that is equivalent to 20% or better reduction in fuel consumption."

With respect to other types of vehicles, Kevin has been involved in just about everything that moves. He began assessing skidoos as early as 1971, motorcycles in 1972, and has over the years, carried out numerous tests on tractor trailers, other heavy-duty transport vehicles and buses, the latter two in the low speed 9m x 9m wind tunnel at Uplands Airport. A recent bus study included the use of computational fluid dynamics as well as wind tunnel testing in the low speed 9m x 9m wind tunnel at Uplands airport. There was a recent study of skidoos to test the engine cooling system and improve rider protection carried out in the 2m x 3m wind tunnel at the J.B. Parkin building, the same facility that was used to test Japanese motorcycles. The results of this study was featured in the September/October 2000 issue of *Cycle Canada.*

One of the questions posed in the article was, "Which of two bikes had the most slippery shape?" It was raised because the one with the larger engine had reached 299.5 km/h while the motorcycle with a smaller engine had reached 305 km/h at tests carried out at the Transport Canada speed oval. It was rumoured the fault lay in the electronic speed control but the wind tunnel provided a simple answer, the article explained. What limited the top speed of the slower vehicle was aerodynamic drag. "Despite the unique monocoque frame design, the expertise of the company's aerospace division and various winglets and spoilers, it was found the more powerful model produced significantly more drag."

It is also, in Kevin's opinion, "a misconception that a sharp projectile-like nose produces less drag; the shape at the back of the object is also a major factor." He adds: "Of course, the biggest single factor producing drag on motorcycles and bicycles is the rider. Being in a crouched position is essential (being small can also produce a 15 per cent reduction) and wearing tight clothing, which reduces the balloon effect, can provide another

A model of the Baltic Millennium Bridge underwent tests in the boundary layer wind tunnel at the University of Western Ontario before the actual bridge was built between Newcastle upon Tyne and Gateshead, England, to allow cyclists and pedestrians to cross the river Tyne. It is one of the hundreds of bridges and buildings tested at the university since the opening of its first boundary layer wind tunnel in the 1960s.

15 per cent reduction," the *Cycle Canada* article reported. The article also points out that bikes meant for racing "typically have enclosed shapes or bodywork that is integrated with the rider's body to produce a smaller wake," and adds, "For street bikes, however, styling conventions and comfort demands make this sort of approach impractical."

Being impractical is certainly not the object at the University of Western Ontario or RWDI who provide roughly 60 per cent the world's "wind engineering" tests on buildings and bridges in their separate boundary layer wind tunnels, located roughly 70 miles apart at London and Guelph, Ontario, respectively.

The words "boundary layer" are used to describe the wind turbulence from ground level to heights of roughly 1500 feet which is modelled in these specialized wind tunnels. As structures, they are comparatively simple when compared with the aerodynamic, climatic or the acoustic wind tunnels designed by Aiolos and the earlier company, DSMA, as the speeds required are roughly 40 to 100km/h and the buildings which house them are also simple. They are, in effect, long box-like structures without complex hot and cold chambers and huge fans or compressed air tanks that are required for testing aircraft and automobiles. They are long in order to allow a deep boundary layer to develop to simulate the full-scale wind. After 80 feet, the boundary layer is about 1/400 of full scale. A city block of houses is represented, therefore, by three to four inch high squares while skyscrapers and other tall structures, such as the CN Tower in downtown Toronto, consist of models that stand between 2 and 3½ feet tall.

Set up at one end of the tunnel, they look like a giant playhouse for children yet they play a key role for business and professional people such as architects, structural designers, civil engineers, bridge builders and contractors. Everything from the foundation footings of bridges and buildings to the design and thickness of the glass required on office and apartment structures have been assessed in these wind tunnels. They generate the wind turbulence and velocities not only at the penthouse level to assess what the potential sway factor might be, but also the effect of the wind at ground level. This is because the geometry of a new building and its location with respect to neighbouring buildings can be an important factor in wind turbulence. Besides that, they have been used to measure the effect of snow drifting and the potential for major damage or disaster if roofs are not properly designed. Boundary layer wind tunnels have also determined that not only wind but torrential rain running down the cables of a bridge structure can contribute to its sway factor. The skin surface or cladding on buildings, roofing materials, window frames and the thickness of the glass in them are major aspects that have been determined by both the boundary layer wind tunnel at the University of Western Ontario and RWDI. Both have garnered solid reputations for such studies all around the world.

The pioneer work on boundary layer wind tunnels began in Canada with Dr. Alan G. Davenport, a native of South Africa who, after receiving his early schooling in that country, obtained his B.A. and M.A. in mechanical engineering at Cambridge University, England, came to the University of Toronto to take his M.A.Sc. in Civil Engineering and then returned to the University of Bristol for a Ph.D. in Civil Engineering in 1961.

Appointed to the Engineering Faculty at the University of Western Ontario that year, Dr. Davenport was soon engaged by the engineers of the World Trade Center in New York City to determine the effect the wind might have on models of its proposed unique twin tower structure. Lacking facilities in Canada, he carried out tests in both Great Britain and in a wind tunnel at Colorado State University that had been funded by the U.S. Army. By 1964, he had also obtained some funds from Canada's National Science and Engineering Research Council to build one of the world's first wind tunnels for the study of civil engineering structures Dr. Davenport designed it and a contract amounting to $30,000 built the structure on the campus in 1965. Shortly afterward, he was engaged to do a study of the Sears Tower in Chicago and, since then, Dr. Davenport and his staff of four senior professors and roughly 25 postgraduate students and staff have tested hundreds of buildings and bridges around the world.

They have also been called on to give evidence in numerous cases where failures have occurred. Dr. Nicholas Isyumov, a faculty member since 1965 and now full professor and research director, besides directing or co-directing wind-engineering studies on hundreds of buildings worldwide, has also been an expert witness in more than a dozen litigations where structural collapses have occurred because of wind or snow factors. Another senior staff member is Dr. David Surry, a graduate of the University of Toronto who joined the staff in 1971. Besides work on such projects as the Hong Kong Bank and the Bank of China in Hong Kong, he has conducted research studies and written papers on low-rise building loads that have been adopted by the National Building Code of Canada and by a number of other countries as well. Other senior members are J. Peter C. King, who has specialized in wind engineering of bridges built in the United States, Norway, Denmark, Australia, Hong Kong and elsewhere, and Professor B.J. Vickery, Ph.D., a native of Australia who has pioneered modern wind design of chimneys and has used wind tunnels to study wind and wave action on off-shore structures. He has also carried out projects in his home country and served the United Nations in the design of a boundary layer wind tunnel in Hungary and a wind engineering laboratory in India.

In order to continue pioneering research in these fields, the university called for an expansion of its facilities with a second state of the art wind tunnel built in the early 1980s. Once again, Dr. Davenport and his staff largely designed it, with Dr. Vickery and Peter King playing major roles, but they also used the expertise of DSMA International for the superstructure housing the test section, the drive and cooling systems and other aerodynamic features in order to complete the project. Paul Straznicky, now a professor at Carleton University, Ottawa, was the key member of the DSMA International 's staff for the project that included the transfer of the original tunnel into the new building. The new wind tunnel, restructured to serve as both a wind and water test facility, is, according to Dr. Surry, one of the few units in the world that can actually scale the waterway and the wind at the same scale.

Dr. Davenport, still active at the University, has served on numerous government committees and international societies and has won numerous honours for his pioneer work in wind engineering. In addition to twice receiving the Gzowski Medal from the Engineering Institute of Canada, he has won the Izaak Walton Killam Memorial Prize, Canada's most distinguished annual award given in recognition of world-class achievement, and, in 1994, the Canada Gold Medal for Science and Engineering awarded by NSERC. Other countries to have honoured Dr. Davenport include the United States and France. He also won the International Award of Merit in Structural Engineering presented in Copenhagen in 1996, and holds honorary degrees from five Canadian universities as well as the University of Louvain, Belguim, the Technical University of Denmark, La Plata University of Argentina and the University of Bristol, U.K.

It was another professor, Frank Theakston, at the University of Guelph, who, when he wanted to determine aspects of drifting snow, co-founded a firm that eventually became RWDI in 1972. William Rowan, a former student of Professor Theakston who had since gone into structural engineering, was the other co-founder. Rowan promoted it into a business enterprise, bringing on board two other associates, Colin Williams and

Anton Davies and, in 1980, added Peter Irwin, a native of England and a Ph.D. graduate in aerodynamics from McGill University who, after working at the National Research Council for a number of years, decided to enter the business world. He became president in 1999, succeeding Rowan who became chairman.

Now a company employing more than 200 people, RWDI is a multi-million dollar enterprise specializing not only in wind tunnel testing but also conducts studies of such things as air pollution dispersion and ventilation requirements in buildings. An even more unique study carried out by them was the trajectory of a baseball hit in a stadium being built for the Texas Rangers. A paper by Michael J. Soligo, a principal of RWDI, reports that such studies started with Comiskey Park when the Chicago White Sox expressed "... an interest in comparing the wind impact on the hitting characteristics between the old stadium and the new baseball park." Since then, similar velocity measurements were done by RWDI for Baltimore, Cleveland, Milwaukee and Seattle. When the study for the Rangers was done for its new ballpark at Arlington, it was discovered that "... an opening behind center field would frequently channel wind into the ballpark and directly into the batter's face. "This, in effect,"the paper added, "would increase the distance a batted ball would travel, as well as create uncomfortable conditions in the seating bowl." Scientifically, Michael explained it in part by writing: "Basically, there are three components of aerodynamic force acting on the baseball – the drag force, the lift force, and the side force – acting lateral to the ball's flight path. The largest aerodynamic force acting on the ball is the drag component. The lift and side forces are caused by asymmetries in the stitch patterns and by the so-called Magnus force, resulting from ball spin." Michael's report added, "The problem was overcome when RWDI designed and tested a unique placement of advertising boards and windscreens."

When the new stadium was built for the Texas Rangers, it was wind tested by RWDI of Guelph, Ontario, who found that wind blowing in from the centre field area definitely affected the flight of the ball and proved a discomfort to fans sitting behind home plate. By recommending changes, the problems were overcome. Photo: Courtesy Rowan, Williams Davis and Irwin.

Baseballs and the stadiums in which it is played are not the only sports facilities to undergo wind-engineering tests by RWDI. Others arenas on its completed project list include hockey arenas like the Air Canada Centre in Toronto, the Kiel Center for the St. Louis Blues, the Marine Midland Arena for the Buffalo Sabres and the MCI Center for the Washington Capitals. Tests were also done on a new stadium for the Cincinnati Bengals football team, and The Fieldhouse, home of the Indiana Pacers basketball team, is another sports facility that was assessed by RWDI.

Not only sports facilities are minutely examined by wind tunnel testing. When Chicago's Orchestra Hall decided to raise the roof, enlarge the stage, extend seating and replace its mechanical, electrical and fire-protection systems, the architects, Skidmore, Owens and Merrill, asked RWDI to conduct a computational fluid design analysis. This was done on a model to identify the acoustical flow patterns because too high velocities could disrupt the sound waves. The study confirmed a complex airflow from the irregular geometry caused by the changes, enabling the architects to adjust the positioning of supply and exhaust ducts as well as the velocity and direction of the air to improve the acoustics of the hall.

Professor Theakston's original concern about snow still remains, however, as one of RWDI's major interest factors, for snow loads on roofs have killed or injured any number of people over the years. One of the most recent tragedies of this kind happened at Sarnia, Ontario, on December 2, 2000, when the roof of a small shopping mall partly collapsed, killing a customer and injuring a clerk. When a major snow storm of roughly 15 inches hit the peninsula area of Washington State in 1996, the roof of a marina collapsed destroying or severely damaging more than 70 luxury yachts and sail boats, with loses estimated in the millions of dollars. Such tragedies and losses might have been avoided if RWDI's Finite Area Element Method, developed in 1986, had been used. It's a computer simulation and wind tunnel combination that measures the effect of drifting snow and snow loads, taking into account such factors as the wind speed and direction, temperature, relative humidity, heat emanating from the building and the development of drifts that can cause unbalanced loads on roofs. With this information, proposals can be made to offset potential problems.

Like the University of Western Ontario, RWDI's major areas of wind engineering studies centre around buildings and bridges. Both establishments have played a major role in making such assessments. A paper on bridge aerodynamics by Gordon Dunn of RWDI indicates, for instance, that the collapse of the long span Tacoma Narrows Bridge in 1940 occurred in winds of just 40 miles an hour because of aerodynamic instability known as flutter. "Wind tunnel testing formed an important part of the investigation into that failure," Gordon wrote, pointing out that "flutter occurs when the wind forces and the dynamic characteristics of the bridge combine in such a way that the wind feeds energy into the oscillations. The bigger the oscillations, the more energy is fed into them and, as there are no limits to these motions, the cycle continues until the structure ultimately fails." He also suggests that baffle plates along the underside of the deck "have become a preferred modification." RWDI finetuned such baffle plates for the Sidney Lanier Bridge in the hurricane area of Georgia to achieve the high flutter speed of 160 mph.

Wind tunnels have also been used to measure the effect of hurricanes on tall buildings, and as more and more office and apartment buildings reach skyward, damping systems have become a more standard means of reducing building or bridge motion. Peter Irwin, as president of RWDI, describes a damping system as an energy-absorbing device to dissipate the energy of the wind. "We've been doing that for about four years now. In fact, for one of our early damper projects we retained Aiolos to assist us in designing dampers in the bridge structure linking the twin Petronas Towers in Kuala Lumpur, " Peter recalls. There are several kinds of dampers. One is basically a pendulum type unit that sways counter to the motion of the structure. This is essentially a Tuned Mass Damper which utilizes a secondary mass attached to a vibrating structure. Another, called a Tuned Liquid Column Damper, is a passive system, the motion in a tank of water acting as a counter movement. RWDI suggests this particular damper can also be used for fire fighting purposes, "... and in some instances can be used for chilled water storage as well." Several other damping techniques are also possible depending on potential wind velocities, structural design and materials used in construction.

The public's growing concern with noise levels and vibration factors has prompted an increasing demand for acoustically sound office and apartment buildings, prompting RWDI to expand their studies and expertise in this area. Car and truck owners are also demanding quieter vehicles as are passengers in buses, trains and planes. As a result, there appears to be a growing need for wind engineering tests and wind tunnel construction in the foreseeable future, but the emphasis for the latter may well change from aerodynamics and climatic to wind tunnels that can assess noise levels.

When it comes to knowing about wind tunnels to test noise levels, Aiolos is ready to provide expertise in that field. Projects cited in this regard include the Honda and Hyundai aeroacoustic wind tunnels and the massive General Motors aerodynamic wind tunnel at Warren, Michigan, where Alistair MacKenzie is project manager for the upgrade to acoustic testing. Earlier, Alistair was project manager of the Large European Acoustic Facility (LEAF), recognized as the world's most sophisticated acoustic chambers to date, that was completed for the European Space Agency (ESA) in 1991 in The Netherlands. It was built to generate noise levels as high as 162dBL in order to replicate noise levels created during lift-off space craft and rockets and to test their effect on satellites.

14 Aiolos Now

Prince Philip visits the Ford Research and Engineering Centre at Dunton, England, in November 2000, where Aiolos Engineering completed a climatic wind tunnel earlier that year. With His Royal Highness are Richard Folkson, senior manager, product verification and testing (left) and Malcolm Thomas, engineering director. Steven Hills, far right, head of the marketing department at Aiolos is in the background with another official of Ford R&D. Photo: Courtesy Ford R&D, Dunton, U.K.

In July 2000, Aiolos Engineering signed a contract to design and oversee the building of a climatic wind tunnel in Vienna that will have the world's longest test section to assess locomotives and coaches manufactured for European railways. In March 2001, Gary Elfstrom and Gerrit Zorgdrager returned from Sweden with an agreement to upgrade the Volvo aerodynamic wind tunnel that must be completed before the end of October. In May, Graham Thomas, as engineering manager of Aiolos, witnessed the signing of a major contract to design and construct a full scale aerodynamic and acoustic wind tunnel and a 40 percent scale model aerodynamic wind tunnel for a consortium made up of two leading French auto makers, Peugeot and Renault, and a third partner, CNAM, a division of the French Ministry of Education. The full-scale wind tunnel will have a nozzle size of 24 square metres. Both wind tunnels will be located at the CNAM site of St. Cyr, situated on the outskirts of Paris. The design work for the wind tunnels will take roughly eight months to complete with the project scheduled to be finished by November 2003.

Graham, a mechanical engineering graduate of Carleton University, worked for Dupont before joining DSMA. He joined Dupont's engineering office in Kingston, left to obtain his Master's degree and rejoined Dupont as a plant engineer at the company's film plant but "didn't feel that being a plant engineer was all I could do." An acquaintance then working for DSMA urged him to send his resume to that company because, "I know you will like it here." He did like it but made one more switch of jobs for seven months before being invited back to DSMA by Gerrit Zorgdrager. Graham then followed Gerrit to work in the fan division of TLT Babcock at Akron, Ohio, when DSMA International was sold in 1992. "Some people just don't fit big companies and I guess I'm one of them," he admits.

In his role as engineering manager, Graham put both his engineering and marketing skills together to help win the project in France, and with the contract signed he is now putting together a team of roughly 20 people to complete the job. Some of them are already in place. As manager of contracts at Aiolos, Romanian-born Adrian Hirsch, who has been arranging for companies to work on various aspects of the Austrian wind tunnel, now adds the French and Swedish projects to his list of activities. For France, he will be working closely with a major French contractor, Spie Batignolles, who were partners with Aiolos in securing the French project. Another Aiolos staff member already identified to be project engineer is Dan Jenkins who has just completed that role at the Ford Dunton Environmental Test Laboratories (ETL) at Basildon, U.K. In charge as the aerodynamics and acoustics expert will be senior dynamicist, Peter Waudby-Smith. Since graduating from Queen's University with a B.Sc. in Engineering Physics in 1981 and the University of Michigan a year later with an M.S.E. in Aerospace engineering, Peter has been involved in the commissioning of wind tunnels in Sweden, Korea and Japan. Following the collapse of DSMA International, he spent a year in Germany with TLT before returning to Canada to be among the first people to join Aiolos Engineering in 1994.

Another who will be working on the French wind tunnels is Roy Fedoruk as the mechanical design engineer. Roy, a native of Toronto, graduated from Ryerson as a mechanical technician and joined DSMA in 1964. Three years later, he was sent to India for two years to work, first, at the blowdown wind tunnel site in Bangalore and then for six months at the Tata-DSMA office in Bombay. Since then he has done design work on more than 30 wind tunnels of every description: blowdown, low speed, climatic, acoustic, icing, plasma and the high speed indraft wind tunnel built at Valcartier, Quebec.

Meanwhile a team of roughly a dozen people have produced literally thousands of detailed design drawings of the various components required for the railroad wind tunnel in Vienna. These include designing the test section, the fan, nozzle, diffusers, compressor, electrical controls and computer software, all under the guidance of former DSMA employees who now hold executive positions with Aiolos. They include Gerrit Zorgdrager as an overseer and Carl Balkwill to supervise the design and commissioning of the numerous controls required for the project.

Besides these latest contract agreements, the company has, since its first full year of operation as Aiolos Engineering Corporation in 1994, completed or been involved on three continents in a number of wind tunnel and other test facilities: General Motors in Michigan, Ford in the U.S.A. and the U.K., Hyundai in Korea, and Williams Grand Prix Engineering Limited of Great Britain for the relocation of a 2.7m x 2.5m aerodynamic wind tunnel (designed by DSMA) to test Formula 1 racing cars at 50 percent scale. Its general manager David Williams wrote in September 1997, "Overall, I am very pleased to say Aiolos carried out the project to spec., on time and to budget. In the Formula One business, this is extremely important particularly regarding Wind Tunnel work as this is our most important design tool."

A month earlier, another letter of commendation for the team who worked on a low speed wind tunnel was received from the Agency for Defense Development of Korea, whose leader, SeungKi Ahn, Ph.D., wrote, "Its aeroacoustic design feature might be the best design ever practiced for the aerospace industry around the world."

Another DSMA veteran deeply involved in the aerodynamic requirements of the Austrian wind tunnel at Vienna is Sergio Raimondo, one of several Aiolos senior aerodynamicists. Early in his career, he was given the challenge of researching the use of slotted wall wind tunnels to test automobiles. By the end of 1994, he was among the dozen former employees of DSMA International or ATCON on the Aiolos staff and which, today, has grown to more than double that number. This includes Gerrit Zorgdrager, Graham Thomas, Alistair MacKenzie, Carmen Volpe, Ratko Dzoja, Ervin Kass, Michael Higgins, Tony Ferreira, and John Seaman of the mechanical design staff, and Warren Letki, Kochi Nishiwaki, and Gurmit Saini in the electrical controls group, and others who have joined since 1995.

The company, however, is not just composed of DSMA veterans. In the last five years a number of men and women have joined the organization including Peter Thuerig, a native of Switzerland and graduate in mechanical engineering at the State College of Engineering at Lucerne. Arriving in Canada in 1978, he worked for several engineering firms in Calgary before moving to Toronto as Manager of Engineering for the Sulzer Technology Corporation between 1985 and 1998. He then joined Aiolos as a senior project manager and, in that capacity, worked on the Sanden International climate wind tunnel at Bad Nauheim, Germany, before switching to a similar role for the railway wind tunnel contract in Vienna. Another is John Kennedy who, after graduating in mechanical engineering from Memorial University at St. John's, Newfoundland, decided to move to Toronto and joined Aiolos as a designer. Later, he went to Korea to work for six months on the ADD low-speed wind tunnel, then spent nearly a year at the Volvo Truck climate wind tunnel site in Sweden and expects to return there later this year to work on the upgrading of the Volvo Car wind tunnel built in the 1980s. With him for six weeks in Sweden was Trevor Bender, with a Master's degree in Industrial aerodynamics from the University of Saskatchewan. In his four years with the company, he has visited Holland for a month to help train staff for the Korean wind tunnel completed in 1999, and was then involved in the commissioning exercises for the Lola wind tunnel in the U.K.

Working in the marketing department with Steve Hills and Gary Elfstrom is Eric Wilson, a Ryerson graduate who also joined the company four years ago. Another is Markus Hess who, before joining Aiolos, had a small company of his own, but "with four children approaching university age, decided he wanted to rely on an assured paycheck rather than an uncertain income." Another recent arrival to join the technical or professional staff is Marta Cavar, a native of Yugoslavia who is working with David Van Every on the research being carried out at NRC. All of them comment positively on the relaxed but professional atmosphere that permeates the office environment and the encouragement they receive to "try things" rather than fall back on routine solutions.

Since the signing of the climatic wind tunnel in Austria, the Sanden International 6 square metre hot/cold climatic wind tunnel at Bad Nauheim, Germany, was commissioned and officially opened in September 2000. Another commissioning completed in May 2001 was that of the 1.5m X2m low-speed wind tunnel at the University of Malaysia where Ratko Dzoja was site manager from 1999 until its completion in 2001. A third project to be completed in 2001 is the conversion of a General Motors aerodynamic wind tunnel at Warren, Michigan, which was built in 1975 into an aeroacoustic facility. The project manager for this job is Alistair MacKenzie, a native of the Highlands of Scotland who worked for DSMA International between 1979 and 1990, spent the next few years with another former DSMA employee, John Farrell, at McMaster University, and returned to Aiolos in 1995. He and his wife

Sandra have since then spent time in Germany and in Seoul during the construction of a 28 square metre aeroacoustic wind tunnel for the Hyundai Motor Company.

Alistair's reports back to Toronto not only include the technical aspects of a job, but the social life of the area. His report from Germany, for instance, explained not only the meaning of the word "bad" in Bad Nauheim but that the city, some 50 km north of Frankfurt, is built around thermal springs, and that the Romans were the first to build baths there, "which are now contained in a very elegant renaissance-style building."

In his technical report he wrote, "We will be working with some of our long-time sub-contractors such as Voith Howden GmbH for the fan and drive system, and new sub-contractors such as Sulzer Escher Wiese for the heating/cooling system, Venti-Oelde GmbH for the shell, and 4-H-Jena Engineering GmbH for the electrical controls. Our subsidiary, Dynamotive Limited, will supply the dynamometer and Facility Control System and the solar simulation will be designed around solar lamps provided by Vortek Industries Limited of Vancouver, B.C."

While in Seoul, he wrote about the Korean favourite pastime of hiking on mountain trails and how his "... local mountain (really a hill) is called Umyonsan which means Woman Mountain – a 15-minute walk to the base from our apartment. At 550m high, 1500m to the top, it takes about two hours to climb depending on the number of rests one has to take." He goes on to explain that many of the trails have exercise benches and some have elaborate exercise machines, and says some city dwellers in Seoul climb the mountain every day before going to work. He also commented on how an 82-year-old Korean musician played folk music. "When I hear the wailing sound of Korean folk music, I could be persuaded that bagpipes have come from the same source, and when you add this to the traditional Korean drumming, the only thing missing are the kilts."

Overseeing all this activity from the launch of Aiolos Engineering in January 1994 to June 2001, was Roland Vander Straeten, the Belgian student who came to Canada in 1970 to take a postgraduate degree in aerodynamics at the University of Toronto. After several jobs and securing an MBA degree from York University in 1978, he joined DSMA International and soon became a senior project manager in both Europe and Asia before leaving the company in 1989. By keeping in touch with a number of employees and contacts he had made on two continents, he was able to form Aiolos in name only in 1992 and Aiolos Engineering as an active company in 1994. Deciding to leave after seven years as president, Roland, it is recalled by other senior members of Aiolos, was the driving force in getting the company launched and guided it to its present-day achievements. Gerrit Zorgdrager, vice president and general manager for the past two years, now overseeing the day to day operations of the company, summed up Roland's contribution by claiming, "He was responsible for bringing us together following the collapse of DSMA International and its sale, first to TLT-Babcock and then to Ransco. He did a tremendous job in getting us some of our first contracts and attracting former key people at DSMA International and ATCON to return, as well as adding some bright new members to the staff. We'll miss him, but we feel we have the skilled people, to not only carry on, but also take on any new challenges."

"There are many kinds of new challenges," says Adrian Hirsch, Manager of Projects for Aiolos. Besides the role of continuing to compete for projects at home and abroad, he is well aware that "as an international consulting firm, banks worry that you don't have a huge number of assets in terms of buildings and equipment. There are also other bank issues, such as guarantees at the national and international level; the exchange rates and inflation or recession swings not only in your own country but also in the countries where you are carrying out contracts."

The key person in dealing with such monetary issues on the Aiolos management team is J.C. Irani, born in India and a CA graduate. He also took additional degrees in England and worked there before moving to Canada where he joined Goldfarb Consultants, a chartered accountant firm in Toronto. As an early assignment with Goldfarb, "J.C." was to upgrade the Aiolos accounting system when it started getting contracts in Asia and Europe. "Roland realized the need to know more about bank issues such as guarantees at the national and international level, or exchange rates and inflation or recession swings not only in your own country but also in the countries were you have contracts, so he invited me to join the team."

As chief of administration and accounting, "J.C." not only oversees the office staff and computing system, but also plays a key role in assessing foreign currency fluctuations. Pointing out that the company's operating expense is essentially in Canadian dollars, but supplies are more often priced in U.S. dollars, he says, "One must constantly keep an eye on how these two currencies interact with other major currencies around the world. It's fascinating," he adds, "particularly when dealing with European and Asian contracts." He recalls when the Euro was almost at par with the U.S. dollar, but then slipped to 80 cents U.S., and an even more dramatic downturn of the Korean economy in December 1998, when that country's currency dropped from 758 won to the U.S. dollar to 2000 won to the U.S. dollar. "Because of these possibilities, when we sign a contract we try to match the currency that we are going to get with the currency that we are going to pay out and any difference we try to get up front (or at least most of it) with a decent down payment, so we have the Canadian dollars to fund our design staff expenses over here," he adds.

J.C. is also quick to point out that there is also the challenge of having a small company like Aiolos compete against a big U.S. corporation which has a lot of resources at its disposal. "Aiolos cannot come anywhere close to commanding those types of resources so the challenge is in managing them in the most efficient manner," he admits. "However," he adds, "in the past few years we have added two subsidiary companies, one in England called Dynamotive PLC and another in Seoul, called Aiolos Korea."

Dynamotive is the larger of the two, having some 20 on staff at Loughborough, U.K. It was started when Clive Barron, as general manager of the dynamometer division of the Brush Electric Machine Company, was a supplier for the British Leyland contract in the early 1980s, when Roland was site manager there. When Brush Electric was sold in 1995, Aiolos financed Clive to set up a company to oversee its growing number of European projects as well as manufacture dynamometers and provide a computer control software staff to offer a basic package suitable for most wind tunnel control rooms. However, when clients want a state-of-the-art computer software program for their facilities, Arthur Ho at ReACT Technologies, in Toronto, is invariably invited to take part in providing such expertise.

Aiolos Korea was created in 1996, shortly after a contract was negotiated for a 28 square metre aeroacoustic wind tunnel for Hyundai Motor Company. It was decided to have Aiolos set up its own agency to not only carry out a marketing function, but also help Aiolos staff members who were sent to Korea to work on the project. Ben Kwon was hired from the Samsung Corporation as general manager. While a truck/bus climatic wind tunnel contract for Hyundai was cancelled following the collapse of the Korean economy in 1998, Aiolos Korea and its small staff of four were instrumental in getting the engine test facilities for the Korean Air Force. "The Air Force was building a trainer and included in it was an engine from Pratt and Whitney Canada," Roland explains, "and they wanted to have a facility to test the engine for maintenance purposes."

As mentioned, Aiolos is also involved in two research projects with the NRC in Ottawa and is also, like NRC and RWDI, now deeply involved in using Computational Fluid Dynamics (CFD), a sophisticated numerical modeling approach for simulating two or three dimensional fluid flows on powerful computers. In further describing CFD, a company project sheet mentions that it can be used "... to solve flow problems involving various gases, liquids and particles. By using CFD," it adds, "it is now possible to model problems which would not historically have been practical to analyse," and suggests, "... the range of applications include such things as rain impact to snow filtration, aerodynamics, industrial processes, combustion, and ventilation."

Another technological boost is the use of Parametric 3D design computers to enable staff and clients to not only see the full design in three dimensions, but also allows information to be more quickly and accurately relayed between various members of the mechanical, electrical or control system design teams.

Moving to new office facilities at 2150 Islington Avenue, Toronto, over the March 31 weekend of 2001, the Aiolos management team and staff of some 50 employees look out from their ground floor offices, conscious that they have, in the past seven years, re-established Canada's reputation as designers and builders of wind tunnels. They look forward to expanding their consulting services and expertise into other areas where a vast knowledge of aerodynamic technology, mechanical design, electrical expertise and computer software skills will be utilized, not only to continue producing sophisticated wind tunnels, but also expand into other areas of engineering consulting services.

Harvey Griggs of H.G. Engineering, who played a major role in getting Aiolos launched and remains one of its key directors, succinctly sums up his view of the company he helped create. "I am impressed with the way this group of engineers, with such diverse backgrounds, came together to establish Aiolos Engineering and recapture Canada's reputation as a leader in wind tunnel design and construction. Like its predecessors, DSMA International and DSMA ATCON, Aiolos has built an enviable worldwide reputation and is truly a unique Canadian success story."